Mathematical Foundations of Time Series Analysis

Jan Beran

Mathematical Foundations of Time Series Analysis

A Concise Introduction

Springer

Jan Beran
Department of Mathematics and Statistics
University of Konstanz
Konstanz, Germany

ISBN 978-3-030-08975-7 ISBN 978-3-319-74380-6 (eBook)
https://doi.org/10.1007/978-3-319-74380-6

Mathematics Subject Classification (2010): 62Mxx, 62M10

Softcover re-print of the Hardcover 1st edition 2017

Printed on acid-free paper

This Springer imprint is published by the registered company Springer International Publishing AG part of Springer Nature.
The registered company address is: Gewerbestrasse 11, 6330 Cham, Switzerland

Preface

The historical development of time series analysis can be traced back to many applied sciences, including economics, meteorology, physics or communications engineering. Theoretical developments of the subject are closely linked to progress in the mathematical theory of stochastic processes and mathematical statistics. There are a number of excellent books on time series analysis, including Grenander and Rosenblatt (1957), Box and Jenkins (1970), Hannan (1970), Anderson (1971), Koopmans (1974), Fuller (1976), Priestley (1981), Brockwell and Davis (1991), Hamilton (1994), Diggle (1996), Brillinger (2001), Chatfield (2003), Lütkepohl (2006), Durbin and Koopmann (2012), Woodward et al. (2016), and Shumway and Stoffer (2017).

Time series analysis is now a well-established scientific discipline with rigorous mathematical foundations. On the other hand, it is a very broad subject area, and, due to the diverse sciences that contributed to its development, the time series vocabulary is permeated with terminology reflecting the diversity of applications (cf. Priestley 1981, Preface p. vii). This book is an attempt to summarize some of the main principles of time series analysis, with the hope that the concise presentation is helpful for teaching students with a mathematical background. The book grew out of lectures taught to students of mathematics, mathematical finance, physics and economics at the University of Konstanz.

I would like to thank Martin Schützner, Mark Heiler, Dieter Schell, Evgeni Shumm, Nadja Schumm, Arno Weiershäuser, Dirk Ocker, Karim Djaidja, Haiyan Liu, Britta Steffens, Klaus Telkmann, Yuanhua Feng, Philipp Sibbertsen, Bikramjit Das, Rafal Kulik, Liudas Giraitis, Sucharita Ghosh and other colleagues for fruitful collaboration; and to Volker Bürkel for reading parts of a preliminary manuscript. Thanks go also to the University of Konstanz for granting me a sabbatical with the purpose of working on this book. Most importantly, I would like to thank my family, Céline, Sucharita and Sir Hastings—our Coton de Tuléar—for keeping me motivated.

Konstanz, Germany — Jan Beran
November 2017

Contents

Chapter 1
Introduction

1.1 What Is a Time Series?

Definition 1.1 Let $k \in \mathbb{N}$, $T \subseteq \mathbb{R}$. A function

$$x : T \to \mathbb{R}^k, t \to x_t$$

or, equivalently, a set of indexed elements of $\mathbb{R}^k$,

$$\left\{x_t | x_t \in \mathbb{R}^k, t \in T\right\}$$

is called an observed time series. We also write

$$x_t \ (t \in T) \text{ or } (x_t)_{t \in T}.$$

Definition 1.2 Let $k \in \mathbb{N}$, $T \subseteq \mathbb{R}$,

$$\Omega = \left(\mathbb{R}^k\right)^T = \text{space of functions } X : T \to \mathbb{R}^k,$$

$$\mathscr{F} = \sigma\text{-algebra on } \Omega,$$

$$P = \text{probability measure on } (\Omega, \mathscr{F}).$$

The probability space $(\Omega, \mathscr{F}, P)$, or equivalently the set of indexed random variables

$$\left\{X_t | X_t \in \mathbb{R}^k, t \in T\right\}, \ (X_t)_{t \in T} \sim P$$

J. Beran, *Mathematical Foundations of Time Series Analysis*,
https://doi.org/10.1007/978-3-319-74380-6_1

Table 1.1 Types of time series $X_t \in \mathbb{R}^k$ $(t \in T)$

Property	Terminology
$k = 1$	Univariate time series
$k \geq 2$	Multivariate time series
T countable, $\forall a < b \in \mathbb{R} : T \cap [a, b]$ finite	Discrete time
T discrete, $\exists u \in \mathbb{R}_+$ s.t. $t_{j+1} - t_j = u$	Equidistant time
$T = [a, b]$ $(a < b \in \mathbb{R})$, $T = \mathbb{R}_+$ or $T = \mathbb{R}$	Continuous time

is called a time series, or time series model. Instead of $(\Omega, \mathcal{F}, P)$ we also write

$$X_t \ (t \in T) \text{ or } (X_t)_{t \in T} .$$

Moreover, for a specific realization $\omega \in \Omega$, we write $X_t(\omega)$ and

$$(x_t)_{t \in T} = (X_t(\omega))_{t \in T} = \text{sample path of } (X_t)_{t \in T} ,$$

$$(x_{t_i})_{i=1,\ldots,n} = (X_{t_i}(\omega))_{i=1,\ldots,n} = \text{finite sample path of } X_t .$$

Remark 1.1 Ω may be more general than in Definition 1.2. Similarly, the index set T may be more general than a subset of $\mathbb{R}$, but it must be ordered and metric. Thus, $(X_t)_{t \in T}$ is a stochastic process with an ordered metric index set T.

Remark 1.2 An overview of the most common types of time series $X_t \in \mathbb{R}^k$ $(t \in T$, $T \neq \emptyset)$ is given in Table 1.1.

Remark 1.3 If X_t =equidistant time series, then we may set w.l.o.g. $T \subseteq \mathbb{Z}$.

1.2 Time Series Versus iid Data

What distinguishes statistical analysis of iid data from time series analysis? We illustrate the question by considering the case of equidistant univariate real valued time series $X_t \in \mathbb{R}$ $(t \in \mathbb{Z})$.

Problem 1.1 Is consistent estimation of P possible?

Solution 1.1 The answer depends on available a priori information and assumptions one is willing to make. This is illustrated in the following.

Notation 1

$$F_{X_t}(x) = P(X_t \leq x) = \textit{marginal distribution function of } X_t \textit{ at time } t$$

$$F_n(x) = \frac{1}{n}\sum_{t=1}^{n} 1\{x_t \leq x\} = \textit{empirical marginal distribution function}$$

Lemma 1.1 *Assumption:*

$$X_t \in \mathbb{R}\ (t \in \mathbb{Z})\ iid.$$

Then

$$\forall t \in \mathbb{Z}: F_{X_t} = F_{X_0}$$

and

$$P \textit{ is fully specified by } F_{X_0}.$$

Theorem 1.1 (Glivenko-Cantelli-Theorem) *Assumption:*

$$X_t \in \mathbb{R}\ (t \in \mathbb{Z})\ iid.$$

Then

$$P\left(\lim_{n\to\infty} \sup_{x\in\mathbb{R}} |F_n(x) - F_{X_0}(x)| = 0\right) = 1.$$

Proof See e.g. van der Vaart (1998). For the original proofs see Glivenko (1933) and Cantelli (1933).

Conclusion 1 *Given the iid assumption, P can be estimated consistently. No additional assumptions are required.*

This is not true in general under non-iid assumptions.

Example 1.1

$$X_t = \theta_t + Z_t\ (t \in \mathbb{Z}),\ Z_t \text{ iid } N(0,1)$$

$$\theta_t\ (t \in \mathbb{Z}) \text{ unknown.}$$

Consistent estimation of θ_t is not possible, unless additional assumptions on θ_t are imposed.

Example 1.2

$$X_t = U \ (t \in \mathbb{Z}),$$

$$0 < p = P(U = 1) = 1 - P(U = 0) < 1.$$

Consistent estimation of p is not possible.

Conclusion 2 *In general, consistent estimation of P is not possible, unless additional assumptions are imposed. The problem is:*

$$\begin{aligned}
\textit{observed time series} &= (x_1, \ldots, x_n) \\
&= (X_1(\omega), \ldots, X_n(\omega)) \\
&= \textit{sample of size one from n-dimensional distribution.} \\
&\neq \textit{sample from the infinite dimensional distribution } P \textit{ on } \mathbb{R}^{\mathbb{Z}}
\end{aligned}$$

Chapter 2
Typical Assumptions

In general, consistent estimation of P is not possible without additional assumptions. In this chapter, typical assumptions used in time series analysis are discussed. For simplicity we focus on equidistant univariate real valued time series $X_t \in \mathbb{R}$ $(t \in \mathbb{Z})$.

2.1 Fundamental Properties

2.1.1 Ergodic Property with a Constant Limit

The asymptotic distribution of many statistics follows—sometimes after a complicated proof or suitable transformations—from the asymptotic distribution of sums.

Notation 2

$$\bar{x} = \bar{x}_n = n^{-1} \sum_{t=1}^{n} X_t$$

Definition 2.1 $X_t \in \mathbb{R}$ $(t \in \mathbb{Z})$ has the almost sure ergodic property with a constant limit, or a.s. EPCL, if

$$\exists \mu \in \mathbb{R} \text{ s.t. } P\left(\lim_{n \to \infty} \bar{x} = \mu\right) = 1.$$

Sometimes one also calls this the mean-ergodic property in the a.s. sense.

J. Beran, *Mathematical Foundations of Time Series Analysis*,
https://doi.org/10.1007/978-3-319-74380-6_2

Definition 2.2 $X_t \in \mathbb{R}$ $(t \in \mathbb{Z})$ has the L^2-ergodic property with a constant limit, or L^2-EPCL, if

$$\exists \mu \in \mathbb{R} \text{ s.t. } \lim_{n\to\infty} E\left[(\bar{x}-\mu)^2\right] = 0.$$

Sometimes one also calls this the mean-ergodic property in the L^2-sense.

Under which circumstances can it happen that the EPCL does not hold? Three main problems can occur, as outlined in the following.

Problem E1 Lack of stability: Distribution of $X_{t+1}, X_{t+2}, \ldots, X_{t+n}$ changes too much as a function of t so that $(x_t)_{t=1,\ldots,n}$ is not sufficiently representative for P on $\mathbb{R}^{\mathbb{Z}}$.

Example 2.1

$$\varepsilon_t \text{ iid}, E(\varepsilon_t) = 0, \sigma_\varepsilon^2 = var(\varepsilon_t) < \infty,$$

$$X_t = \beta t + \varepsilon_t.$$

Then

$$\bar{x} = \beta \frac{1}{n} \frac{n(n+1)}{2} + \bar{\varepsilon},$$

$$P(|\bar{x}| \to \infty) = 1.$$

Example 2.2

$$\varepsilon_t \text{ iid}, E(\varepsilon_t) = 0, \sigma_\varepsilon^2 < \infty,$$

$$X_s = 0 \ (s \le 0), X_t = \sum_{s=1}^{t} \varepsilon_s \ (t \ge 1).$$

Then

$$\bar{x} = n^{-1} \sum_{t=1}^{n} t\varepsilon_{n-t+1},$$

$$var(\bar{x}) = \frac{\sigma_\varepsilon^2}{n^2} \sum_{t=1}^{n} t^2 \to \infty.$$

Problem E2 High variability of the marginal distribution F_{X_t}.

Example 2.3 For Cauchy distributed iid X_t, we have

$$\bar{x} \underset{d}{=} X_1.$$

Problem E3 Absorbing states:

$$X_t \ (t \in \mathbb{Z}) \ \mathscr{F}_t\text{-measurable},$$

$$\exists t \in \mathbb{Z}, A_t \in \mathscr{F}_t \text{ s.t. } 0 < P(X_t \in A_t) < 1 \text{ and } P(\forall s > t : X_s \in A_t | A_t) = 1.$$

Then

$$A_t = \text{absorbing state}.$$

Example 2.4

$$X_t = U \ (t \in \mathbb{Z}),$$

$$0 < p = P(U = 1) = 1 - P(U = 0) < 1.$$

Then

$$A_t = \{X_t = 1\} = \text{absorbing state}.$$

2.1.2 *Strict Stationarity*

Definition 2.3 $X_t \in \mathbb{R}$ $(t \in \mathbb{Z})$ is called strictly stationary or strongly stationary, if

$$\forall k \in \mathbb{Z},\ \forall m \in \mathbb{N},\ \forall t_1, \ldots, t_m \in \mathbb{Z}\colon\ (X_{t_1}, \ldots, X_{t_m}) \underset{d}{=} (X_{t_1+k}, \ldots, X_{t_m+k}).$$

Example 2.5 $X_t \in \mathbb{R}$ $(t \in \mathbb{Z})$ iid is strictly stationary.

Example 2.6

$$X_t = \sum_{j=0}^{q} \psi_j \varepsilon_{t-1} \ (t \in \mathbb{Z}),$$

$$\varepsilon_t \in \mathbb{R} \ (t \in \mathbb{Z}) \text{ iid}, \ \psi_j \in \mathbb{R} \ (j = 0, \ldots, q)$$

is strictly stationary. X_t is called a moving average process of order q, or MA(q) process.

Remark 2.1 Strict stationarity solves Problem E1, but not Problems E2 and E3.

2.1.3 Weak Stationarity

Definition 2.4 Let

$$X_t \in \mathbb{R}\ (t \in \mathbb{Z}) \text{ s.t. } \forall t \in \mathbb{Z}\colon E\left(|X_t|\right) < \infty.$$

Then

$$\mu_t = E\left(X_t\right)\ \ (t \in \mathbb{Z})$$

is called the expected value function, or mean function, of X_t. If

$$E\left(X_t^2\right) < \infty,$$

then

$$\gamma : \mathbb{Z} \times \mathbb{Z} \to \mathbb{R}$$

with

$$\gamma\left(s,t\right) = cov\left(X_s, X_t\right) = E\left[\left(X_s - \mu_s\right)\left(X_t - \mu_t\right)\right]$$

is called the autocovariance function (acf) of X_t, and

$$\rho\left(s,t\right) = corr\left(X_t, X_s\right) = \frac{\gamma\left(s,t\right)}{\sqrt{\gamma\left(s,s\right)\gamma\left(t,t\right)}}$$

is called the autocorrelation function (acf) of X_t.

Remark 2.2

$$\gamma\left(t,t\right) = var\left(X_t\right), \rho\left(t,t\right) = 1$$

Remark 2.3 For

$$X_t \in \mathbb{C}\ (t \in \mathbb{Z}),$$

we define

$$\gamma\left(s,t\right) = cov\left(X_s, X_t\right) = E\left[\left(X_s - \mu_s\right)\overline{\left(X_t - \mu_t\right)}\right].$$

Lemma 2.1

$$\gamma(t,s) = \overline{\gamma(s,t)}$$

Proof

$$\overline{\gamma(s,t)} = \overline{E\left[(X_s - \mu_s)\overline{(X_t - \mu_t)}\right]} = E\left[\overline{(X_s - \mu_s)\overline{(X_t - \mu_t)}}\right]$$
$$= E\left[(X_t - \mu_t)\overline{(X_s - \mu_s)}\right] = \gamma(t,s)$$

Definition 2.5 $X_t \in \mathbb{R}$ $(t \in \mathbb{Z})$ is called second order stationary, or weakly stationary, if

$$E\left(X_t^2\right) < \infty,$$

$$\exists \mu \in \mathbb{R} \text{ s.t. } \forall t \in \mathbb{Z} : E(X_t) = \mu,$$

$$\exists \gamma : \mathbb{Z} \to \mathbb{R} \text{ s.t. } \forall s,t \in \mathbb{Z} : cov(X_s, X_t) = \gamma(t-s).$$

Lemma 2.2

$$\textit{weak stationarity} \nLeftrightarrow \textit{strong stationarity}$$

Proof Counterexample for $\nRightarrow$:

$$X_{2i} = Z_{2i}\ (i \in \mathbb{Z}), X_{2i+1} = \frac{Z_{2i+1}^2 - 1}{\sqrt{2}}\ (i \in \mathbb{Z})$$

where

$$Z_i\ (i \in \mathbb{Z}) \text{ iid } N(0,1)\text{-variables}$$

Counterexample for $\nLeftarrow$:

$$X_t\ (t \in \mathbb{Z}) \text{ iid, Cauchy distributed}$$

Lemma 2.3 *Assumptions:*

$$X_t \in \mathbb{R}\ (t \in \mathbb{Z}) \textit{ strictly stationary, } E(X_t^2) < \infty$$

Then

$$X_t\ (t \in \mathbb{Z})\ \textit{weakly stationary}$$

Proof

a) μ: The Cauchy-Schwarz inequality implies

$$E^2\left(|X_t|\right) \leq E\left[X_t^2\right] < \infty,$$

and hence

$$\exists \mu_t = E\left(X_t\right) \in \mathbb{R}.$$

Thus, together with strong stationarity

$$\forall s, t \in \mathbb{Z} : \mu_s = \mu_t = \mu \in \mathbb{R}$$

b) γ: The Cauchy-Schwarz inequality implies

$$E^2\left(X_s X_t\right) \leq E^2\left(|X_s|\,|X_t|\right) \leq E\left(X_s^2\right) E\left(X_t^2\right) < \infty.$$

Together with strong stationarity we then have

$$\forall t, k \in \mathbb{Z} : E\left(X_t X_{t+k}\right) = E\left(X_0 X_k\right),$$

and

$$cov\left(X_t, X_{t+k}\right) = cov\left(X_0, X_k\right) = E\left(X_0 X_k\right) - \mu^2 = \gamma\left(k\right).$$

Remark 2.4 Weak stationarity solves Problem E1 w.r.t. first two moments of X_t, and Problem E2 in the sense that $E\left(X_t^2\right) < \infty$. It does not solve Problem E3.

Example 2.7

$$X_t = U\ (t \in \mathbb{Z})$$

where

$$0 < p = P\left(U = 1\right) = 1 - P\left(U = 0\right) < 1,$$

is weakly and strictly stationary, but

$$A_t = \{X_t = 1\} = \text{absorbing state}.$$

2.1.4 Weak Stationarity and Hilbert Spaces

For an introduction to Hilbert spaces see e.g. Young (1988).

Definition 2.6

$$\mathscr{R} = \text{space of } \mathbb{R}\text{-valued random variables on probability space } (\Omega, \mathscr{F}, P),$$

$$L^2(\Omega) = L^2(\Omega, \mathbb{R}) = \left\{X | X \in \mathscr{R} \text{ s.t. } E\left(X^2\right) < \infty\right\},$$

$$\langle X, Y\rangle = \langle X, Y\rangle_{L^2(\Omega)} = E(XY) \;\; (X, Y \in L^2(\Omega)),$$

$$\|X\|^2_{L^2(\Omega)} = \langle X, X\rangle \;\; (X \in L^2(\Omega)).$$

Remark 2.5

$$\|X\|^2_{L^2(\Omega)} = var(X) + E^2(X)$$

Remark 2.6 Analogous definitions for

$$\mathscr{C} = \text{space of } \mathbb{C}\text{-valued random variables on } (\Omega, \mathscr{F}, P),$$

with

$$\langle X, Y\rangle = \langle X, Y\rangle_{L^2(\Omega.\mathbb{C})} = E\left(X\overline{Y}\right).$$

Lemma 2.4 $\langle X, Y\rangle_{L^2(\Omega.\mathbb{R})}$ *and* $\langle X, Y\rangle_{L^2(\Omega.\mathbb{C})}$ *are scalar products, if we interpret* $X = Y$ *as* $P(X = Y) = 1$.

Proof

1)

$$\langle X, Y\rangle = E\left(X\overline{Y}\right) = E\left(\overline{Y}X\right) = E\left(\overline{Y\overline{X}}\right) = \overline{E\left(Y\overline{X}\right)} = \overline{\langle Y, X\rangle}$$

2)

$$\langle aX, Y\rangle = E\left(aX\overline{Y}\right) = aE\left(X\overline{Y}\right) = a\langle X, Y\rangle$$

3)

$$\langle X + Y, Z\rangle = E\left[(X + Y)\overline{Z}\right] = E\left(X\overline{Z}\right) + E\left(X\overline{Z}\right) = \langle X, Z\rangle + \langle Y, Z\rangle$$

4)

$$\langle X, X\rangle = E\left(|X|^2\right) \geq 0$$

$$\langle X, X\rangle = E\left(|X|^2\right) = 0 \Leftrightarrow X = 0 \text{ a.s.}$$

Lemma 2.5 *$L^2(\Omega, \mathbb{R})$ and $L^2(\Omega, \mathbb{C})$ are complete.*

Proof We need to show the implication:

$$X_n\ (n \in \mathbb{N}) \text{ Cauchy sequence in } L^2(\Omega)$$

$$\Downarrow$$

$$\exists \text{random variable } X \text{ s.t. a) } \lim_{n\to\infty} \|X_n - X\|^2 = 0 \text{ and b) } X \in L^2(\Omega).$$

a)

$$X_n \in L^2(\Omega)\ \ (n \in \mathbb{N}) \text{ Cauchy sequence} \Leftrightarrow \|X_n - X_m\| \to 0\ (m, n \to \infty)$$

Hence,

$$\exists \text{subsequence } X_{n_j}\ (j \in \mathbb{N}) \text{ s.t. } \left\|X_{n_{j+1}} - X_{n_j}\right\| \leq 2^{-j}\ (j \in \mathbb{N}).$$

Setting $n_0 = 0, X_0 = 0$, we have

$$X_{n_j} = \sum_{i=1}^{j} \left(X_{n_i} - X_{n_{i-1}}\right).$$

Define

$$W_j = \sum_{i=1}^{j} \left|X_{n_i} - X_{n_{i-1}}\right|.$$

Then

$$\left|X_{n_j}\right| = \left|\sum_{i=1}^{j} \left(X_{n_i} - X_{n_{i-1}}\right)\right| \leq W_j,$$

and therefore

$$E\left(\left|X_{n_j}\right|\right) \leq E\left(W_j\right) = \sum_{i=1}^{j} E\left(\left|X_{n_i} - X_{n_{i-1}}\right|\right).$$

The Cauchy-Schwarz inequality implies

$$E\left(\left|X_{r_i} - X_{n_{i-1}}\right|\right) \leq \left\|X_{n_i} - X_{n_{i-1}}\right\|,$$

so that

$$\begin{aligned} E\left(\left|X_{n_j}\right|\right) \leq W_j &\leq \sum_{i=1}^{j} \left\|X_{n_i} - X_{n_{i-1}}\right\| \\ &\leq |X_1\| + \sum_{i=1}^{j} 2^{-j} \leq \|X_1\| + 1 < \infty. \end{aligned}$$

Since

$$W_j \geq 0 \text{ nondecreasing, } E\left(W_j\right) \leq \|X_1\|_{L^2(\Omega)} + 1 < \infty,$$

we obtain

$$P\left(\exists W = \lim_{j\to\infty} W_j < \infty\right) = 1$$

and hence

$$P\left(\exists X = \lim_{j\to\infty} X_{n_j} \in \mathbb{R}\right) = 1.$$

b)

$$\begin{aligned} \|X_n - X\|^2 = E\left(\left|X_n - \lim_{j\to\infty} X_{n_j}\right|^2\right) &= E\left(\lim_{j\to\infty} \left|X_n - X_{n_j}\right|^2\right) \\ &= E\left(\liminf_{j\to\infty} \left|X_n - X_{n_j}\right|^2\right) \leq \liminf_{j\to\infty} E\left(\left|X_n - X_{n_j}\right|^2\right) \\ &= \liminf_{j\to\infty} \left\|X_n - X_{n_j}\right\|^2. \end{aligned}$$

Now

$$X_n \text{ Cauchy sequence} \Rightarrow \lim_{n\to\infty} \liminf_{j\to\infty} \left\| X_n - X_{n_j} \right\|^2 = 0$$

$$\Rightarrow \lim_{n\to\infty} \|X_n - X\|^2 \le \lim_{n\to\infty} \liminf_{j\to\infty} \left\| X_n - X_{n_j} \right\|^2 = 0.$$

c)

$$\|X\| \le \|X - X_n\| + \|X_n\| < \infty \Rightarrow X \in L^2(\Omega)$$

Corollary 2.1 *$L^2(\Omega, \mathbb{R})$ and $L^2(\Omega, \mathbb{C})$ are Hilbert spaces.*

Proof Follows from Lemma 2.5.

Corollary 2.2 *Assumption:*

$$X_t \in \mathbb{R}\ (t \in \mathbb{Z})\ \text{weakly stationary.}$$

Then

$$\forall t \in \mathbb{Z} : X_t \in L^2(\Omega)$$

Proof Follows from the definition of weak stationarity with $E(X_t^2) < \infty$.

Corollaries 2.1 and 2.2 are useful for linear prediction:

Definition 2.7 Let

$$\mathscr{F}_{\le t} = \sigma(X_s, s \le t),$$

$$\mathscr{X}_t = \left\{ Y \mid Y \in L^2(\Omega), \mathscr{F}_{\le t}\text{-measurable} \right\}.$$

Then, for $k \ge 1$ and $Y \in \mathscr{X}_t$,

$$\sigma_k^2(Y) = E\left[(X_{t+k} - Y)^2\right]$$
$$= k\text{-step prediction mean squared error of } Y$$

Problem 2.1 Find prediction of X_{t+k} given $\mathscr{F}_{\le t}$ that minimizes σ_k^2.

Solution 2.1 Projection theorem for Hilbert spaces.

Theorem 2.1 (Continuity) *Assumptions:*

$$(H, \langle \cdot, \cdot \rangle) \text{ Hilbert space,}$$

$$x, y, x_n, y_n \in H \ (n \in \mathbb{N}) \text{ s.t. } \lim_{n \to \infty} \|x_n - x\| = 0, \ \lim_{n \to \infty} \|y_n - y\| = 0$$

Then

$$\lim_{n \to \infty} \|x_n\| = \|x\|$$

and

$$\lim_{n \to \infty} \langle x_n, y_n \rangle = \langle x, y \rangle .$$

Proof

a)

$$\|x_n\| \le \|x_n - x\| + \|x\|$$

implies

$$\|x_n\| - \|x\| \le \|x_n - x\| .$$

Similarly,

$$\|x\| \le \|x_n - x\| + \|x_n\|$$

implies

$$-\|x_n - x\| \le \|x_n\| - \|x\| .$$

Hence,

$$0 \le \lim_{n \to \infty} |\|x_n\| - \|x\|| \le \lim_{n \to \infty} \|x_n - x\| = 0$$

b)

$$\begin{aligned} \lim_{n \to \infty} |\langle x_n, y_n \rangle - \langle x, y \rangle| &\le \lim_{n \to \infty} (|\langle x_n, y_n - y \rangle| + |\langle x_n - x, y \rangle|) \\ &\le \lim_{n \to \infty} \|x_n\| \|y_n - y\| + \lim_{n \to \infty} \|x_n - x\| \|y\| = 0 \end{aligned}$$

Theorem 2.2 (Projection Theorem) *Assumptions:*

$$(H, \langle \cdot, \cdot \rangle) \text{ Hilbert space, } x \in H,\ A \subseteq H,\ A \text{ closed, convex}$$

Then

$$\exists! \hat{x} \in A \text{ s.t. } \|x - \hat{x}\| = \inf_{y \in A} \|x - y\| \, .$$

Proof

1) Existence: Let

$$d_x = \inf_{y \in A} \|x - y\| \, ,$$

$$y_n \in A \ (n \in \mathbb{N}) \text{ s.t. } \|x - y_n\|^2 < d_x^2 + \frac{1}{n}.$$

Convexity of A implies

$$\frac{y_n + y_m}{2} \in A$$

and therefore

$$\left\| \frac{y_n + y_m}{2} - x \right\| \geq d_x.$$

Now

$$\begin{aligned} \|y_n - y_m\|^2 &= \|(y_n - x) + (x - y_m)\|^2 \\ &= \|y_n - x\|^2 + \|y_m - x\|^2 - 2 \langle y_n - x, y_m - x \rangle \end{aligned}$$

and

$$\begin{aligned} 4 \left\| \frac{y_n + y_m}{2} - x \right\|^2 &= \|(y_n - x) + (y_m - x)\|^2 \\ &= \|y_n - x\|^2 + \|y_m - x\|^2 + 2 \langle y_n - x, y_m - x \rangle \, , \end{aligned}$$

so that

$$4 \left\| \frac{y_n + y_m}{2} - x \right\|^2 + \|y_n - y_m\|^2 = 2 \|y_n - x\|^2 + 2 \|y_m - x\|^2 \, .$$

Hence,

$$\begin{aligned}\|y_n - y_m\|^2 &= 2\|y_n - x\|^2 + 2\|y_m - x\|^2 - 4\left\|\frac{y_n + y_m}{2} - x\right\|^2 \\ &\le 2\|y_n - x\|^2 + 2\|y_m - x\|^2 - 4d_x^2 \\ &\le 2\left(d_x^2 + \frac{1}{n}\right) + 2\left(d_x^2 + \frac{1}{m}\right) - 4d_x^2 = \frac{2}{n} + \frac{2}{m},\end{aligned}$$

so that

$$y_n = \text{Cauchy sequence},$$

$$\exists \hat{x} = \lim_{n\to\infty} y_n \in \bar{A} = A$$

and

$$\|x - y_n\|^2 < d_x^2 + \frac{1}{n}.$$

Thus,

$$\hat{x} \in A \text{ and } \|\hat{x} - y\| = \inf_{y\in A} \|x - y\|.$$

2) Uniqueness: Let

$$\hat{x}_1, \hat{x}_2 \text{ s.t. } \|\hat{x}_1 - y\| = \|\hat{x}_2 - y\| = \inf_{y\in A}\|x - y\| = d_x.$$

Convexity of A implies

$$\frac{\hat{x}_1 + \hat{x}_2}{2} \in A$$

and hence

$$\left\|\frac{\hat{x}_1 + \hat{x}_2}{2} - x\right\|^2 \ge d_x^2.$$

Then

$$\begin{aligned}\|\hat{x}_1 - \hat{x}_2\|^2 &= 2\|\hat{x}_1 - x\|^2 + 2\|\hat{x}_2 - x\|^2 - 4\left\|\frac{\hat{x}_1 + \hat{x}_2}{2} - x\right\|^2 \\ &= 4d_x^2 - 4\left\|\frac{\hat{x}_1 + \hat{x}_2}{2} - x\right\|^2 \\ &\le 4d_x^2 - 4d_x^2 = 0.\end{aligned}$$

so that

$$\|\hat{x}_1 - \hat{x}_2\|^2 = 0.$$

Definition 2.8 Let $(H, \langle\cdot,\cdot\rangle)$ be a Hilbert space over $\mathbb{R}$. Then $A \subseteq H$ is called a linear subspace of H, if

$$0 \in A,$$

$$\forall x, y \in A : x + y \in A,$$

$$\forall x \in A, \lambda \in \mathbb{R} : \lambda x \in A.$$

An analogous definition applies for H over $\mathbb{C}$.

Definition 2.9

$$\begin{aligned} A^{\perp} &= \{x \mid x \in H \text{ s.t. } \forall y \in A : \langle x, y\rangle = 0\} \\ &= \text{orthogonal complement of } A \end{aligned}$$

We write

$$A \perp B \Leftrightarrow \forall x \in A, \forall y \in B : \langle x, y\rangle = 0.$$

Lemma 2.6

$$(H, \langle\cdot,\cdot\rangle) \textit{ Hilbert space over } \mathbb{R} \textit{ or } \mathbb{C}, A \subseteq H \textit{ linear subspace}$$

$$\Rightarrow A \textit{ convex}$$

Proof Let

$$c_1, \ldots, c_k \text{ s.t. } \sum c_i = 1, x_1, \ldots, x_k \in A.$$

Then

$$A \text{ linear subspace} \Rightarrow \sum_{i=1}^{k} c_i x_i \in A.$$

Hence A is convex.

Lemma 2.7

$$(H, \langle\cdot,\cdot\rangle) \textit{ Hilbert space, } A \subseteq H$$

$$\Rightarrow A^{\perp} \textit{ closed linear subspace}$$

Proof

a)

$$\forall x \in A : \langle 0, x \rangle = 0 \Rightarrow 0 \in A^{\perp}$$

b)

$$y, z \in A^{\perp} \Rightarrow \forall x \in A : \langle y + z, x \rangle = \langle y, x \rangle + \langle z, x \rangle = 0 \Rightarrow y + z \in A^{\perp}$$

c)

$$y \in A^{\perp} \Rightarrow \forall x \in A : \langle \lambda y, x \rangle = \lambda \langle y, x \rangle = 0 \Rightarrow \lambda y \in A^{\perp}$$

Thus, a), b), c) imply

$$A = \text{linear subspace.}$$

d) Let

$$y_n \in A^{\perp}, y \in H \text{ s.t. } \lim_{n \to \infty} \|y_n - y\| = 0.$$

Then

$$\forall x \in A : \langle y, x \rangle = \lim_{n \to \infty} \langle y_n, x \rangle = 0 \Rightarrow y \in A^{\perp} \Rightarrow A^{\perp} \text{ closed}$$

Corollary 2.3 *Assumptions:*

$$(H, \langle \cdot, \cdot \rangle) \text{ Hilbert space, } x \in H,\ A = \text{closed linear subspace, } \hat{x} \in A$$

Then

$$\|x - \hat{x}\| = \inf_{y \in A} \|x - y\| \Leftrightarrow x - \hat{x} \in A^{\perp}$$

Proof

1) "$\Leftarrow$" Let

$$\hat{x} \in A \text{ s.t. } x - \hat{x} \in A^{\perp}$$

and

$$y \in A \text{ arbitrary.}$$

Then

$$A \text{ linear space} \Rightarrow \hat{x} - y \in A$$

and

$$x - \hat{x} \in A^{\perp} \Rightarrow \|x - y\|^2 = \|x - \hat{x}\|^2 + \|\hat{x} - y\|^2 \geq \|x - \hat{x}\|^2$$

so that

$$\|x - \hat{x}\|^2 = \inf_{y \in A} \|x - y\|^2 .$$

2) "$\Rightarrow$" Follows from "$\Leftarrow$" and uniqueness.

Definition 2.10 Let A and $\hat{x}$ be as in Corollary 2.3. Then

$$P_A : H \to A, x \to P_A(x) = \hat{x}$$

is called (orthogonal) projection mapping, or projection operator on A.

Definition 2.11 Let $X_t \in \mathbb{R}$ $(t \in \mathbb{Z})$ be weakly stationary,

$$\mathscr{F}_{\leq t} = \sigma(X_s, s \leq t),$$

$$\mathscr{X}_t = \left\{ Y \mid Y \in L^2(\Omega), \mathscr{F}_{\leq t}\text{-measurable} \right\},$$

$$k \in \mathbb{N}, \hat{X}_{t+k} \in \mathscr{X}_t.$$

Then

$$\hat{X}_{t+k} = \text{optimal forecast of } X_{t+k} \text{ given } \mathscr{F}_{\leq t} \text{ (or given } X_s, s \leq t)$$

$$\Leftrightarrow$$

$$\sigma_k^2\left(\hat{X}_{t+k}\right) = \inf_{Y \in \mathscr{X}_t} \sigma_k^2(Y).$$

Definition 2.12 Let

$$(\Omega, \mathscr{F}, P) = \text{probability space},$$

$$X : \Omega \to \mathbb{R} \text{ s.t. } X \text{ is } \mathscr{F}\text{-measurable},$$

$$\mathscr{G} \subseteq \mathscr{F} \text{ s.t. } \mathscr{G} = \sigma\text{-algebra}.$$

A $\mathscr{G}$-measurable random variable Y is called conditional expected value of X given $\mathscr{G}$, if

$$\forall A \in \mathscr{G} : \int_A YdP = \int_A XdP.$$

We then write

$$Y = E\left(X \mid \mathscr{G}\right).$$

Remark 2.7 If $E(X)$ exists, then $E(X|\mathscr{G})$ exists and is unique in the sense of almost sure equality (see e.g. Kallenberg 2002).

Corollary 2.4 *Assumptions:*

$$X_t \in \mathbb{R}\ (t \in \mathbb{Z})\ \textit{weakly stationary},$$

$$\hat{X}_{t+k} \in \mathscr{X}_t.$$

Then

$$\sigma_k^2\left(\hat{X}_{t+k}\right) = \inf_{Y \in \mathscr{X}_t} \sigma_k^2\left(Y\right) \Leftrightarrow \hat{X}_{t+k} \underset{a.s.}{=} E\left(X_{t+k} \mid \mathscr{F}_{\leq t}\right).$$

Proof

$$X_t, X_{t+k} \in L^2\left(\Omega\right),\ \mathscr{X}_t = \text{closed convex subset}$$

Then, for $\hat{X}_{t+k} \in \mathscr{X}_t$,

$$\left\|X_{t+k} - \hat{X}_{t+k}\right\| = \inf_{Y \in \mathscr{X}_t} \left\|X_{t+k} - Y\right\| \Leftrightarrow \hat{X}_{t+k} \in \mathscr{X}_t \text{ and } X_{t+k} - \hat{X}_{t+k} \in \mathscr{X}_t^{\perp}$$

and this random variable is unique. Now

$$\forall Z \in \mathscr{X}_t : E\left[ZE\left(X_{t+k} \mid \mathscr{F}_{\leq t}\right)\right] = E\left[ZX_{t+k}\right]$$

so that

$$\forall Z \in \mathscr{X}_t : \langle Z, X_{t+k} - E\left(X_{t+k} \mid \mathscr{F}_{\leq t}\right)\rangle = 0,$$

and therefore

$$\left\|X_{t-k} - E\left(X_{t+k} \mid \mathscr{F}_{\leq t}\right)\right\| = \inf_{Y \in \mathscr{X}_t} \left\|X_{t+k} - Y\right\|.$$

Problem 2.2 Find a linear prediction of X_{t+k} given $\mathscr{F}_{\le t}$ that minimizes σ_k^2.

Solution 2.2 Orthogonal projection on linear space.

Definition 2.13

$$L_t^0 = \left\{ Y \mid Y = \sum_{j=1}^{k} a_j X_{t_j},\, k \in \mathbb{N},\, a_j \in \mathbb{R},\, t_j \in \mathbb{Z},\, t_j \le t \right\},$$

$$L_t = \overline{L_t^0} = \left\{ Y \mid \exists Y_n \in L_t^0\ (n \in \mathbb{N}) \text{ s.t. } \lim_{n\to\infty} \|Y - Y_t\|^2_{L^2(\Omega)} = 0 \right\},$$

$$L_{-\infty} = \cap_{t=-\infty}^{\infty} L_t = \text{infinite linear past of } X_t$$

Lemma 2.8

$$\forall s \in \mathbb{Z}\text{: } L_{-\infty} = \cap_{t=-\infty}^{s} L_t$$

Proof

$$\forall s \le t\text{: } L_s \subseteq L_t \Rightarrow \forall s \in \mathbb{Z}\text{: } L_{-\infty} = \cap_{t=-\infty}^{s} L_t$$

Definition 2.14 Let $X_t \in \mathbb{R}$ $(t \in \mathbb{Z})$ be weakly stationary, $k \ge 1$ and $\hat{X}_{t+k} \in L_t$. Then

$$\hat{X}_{t+k} = \text{optimal linear forecast of } X_{t+k} \text{ given } \mathscr{F}_{\le t} \text{ (or given } X_s,\, s \le t)$$

$$\Leftrightarrow$$

$$\sigma_k^2\left(\hat{X}_{t+k}\right) = \inf_{Y \in L_t} \sigma_k^2\,(Y).$$

We then write

$$\sigma_{k,\text{opt}}^2 = \sigma_k^2\left(\hat{X}_{t+k}\right).$$

Definition 2.15 Let $X_t \in \mathbb{R}$ $(t \in \mathbb{Z})$ be weakly stationary. Then X_t is called deterministic or singular, if

$$\sigma_{1,\text{opt}}^2 = 0.$$

More generally, a weakly stationary time series Z_t $(t \in \mathbb{Z})$ is called deterministic with respect to X_t $(t \in \mathbb{Z})$, if

$$\forall t \in \mathbb{Z} : \inf_{Y \in L_t} E\left[(Z_{t+1} - Y)^2\right] = 0.$$

Problem 2.3 Existence of optimal linear prediction $\hat{X}_{t+k}$.

Solution 2.3 Existence of orthogonal projection.

Lemma 2.9 *Assumptions:*

$$X_t \in \mathbb{R}\ (t \in \mathbb{Z})\ \textit{weakly stationary},$$

$$\hat{X}_{t+k} \in L_t.$$

Then

$$\sigma^2_{k,opt} < \infty$$

and

$$\sigma^2_k\left(\hat{X}_{t+k}\right) = \sigma^2_{k,opt} \Leftrightarrow \hat{X}_{t+k} = P_{L_t}\left(X_{t+k}\right)$$

Proof

$$L_t = \text{closed linear subspace of } L^2\left(\Omega\right)$$

Hence

$$\sigma^2_k\left(\hat{X}_{t+k}\right) = \sigma^2_{k,\text{opt}} \Leftrightarrow \hat{X}_{t+k} = P_{L_t}\left(X_{t+k}\right).$$

Moreover,

$$\left\|P_{L_t}\left(X_{t+k}\right)\right\|^2 \leq \left\|X_{t+k}\right\|^2 < \infty \Rightarrow \sigma^2_{k,\text{opt}} < \infty.$$

Lemma 2.10 *Assumptions:*

$$X_t \in \mathbb{R}\ (t \in \mathbb{Z})\ \textit{weakly stationary},$$

$$\sigma^2_{1,opt} = 0.$$

Then

$$\forall k \geq 1 : \sigma^2_{k,opt} = 0.$$

Proof

$$\sigma^2_{1,\text{opt}} = 0$$

implies

$$X_{t+k} \underset{a.s.}{=} P_{L_{t+k-1}}(X_{t+k}) \underset{a.s.}{=} P_{L_{t+k-2}}(X_{t+k}) \underset{a.s.}{=} \ldots \underset{a.s.}{=} P_{L_{t+k-j}}(X_{t+k}) \ (j \geq 1).$$

Hence,

$$\sigma^2_{k,\text{opt}} = 0.$$

The calculation of optimal linear forecasts is simplified by the so-called Wold decomposition (Theorem 2.3 below).

Lemma 2.11 *Assumption:*

$$X_t \in \mathbb{R} \ (t \in \mathbb{Z}) \text{ weakly stationary.}$$

Then

$$Z_t \text{ deterministic w.r.t. } X_t \Leftrightarrow \forall t \in \mathbb{Z} : Z_t \in L_{-\infty}$$

Proof

1)

$$\ldots \subseteq L_{t-2} \subseteq L_{t-1} \subseteq L_t$$

implies

$$\forall t \in \mathbb{Z} : L_t = \bigcup_{s=-\infty}^{t} L_s, \ L_{-\infty} = \bigcap_{s=-\infty}^{t} L_s.$$

2) "$\Rightarrow$":

$$Z_t \ (t \in \mathbb{Z}) \text{ deterministic w.r.t. } X_t$$

implies

$$\forall s < t : Z_t = P_{L_s}(X_t) + E_t, \ E_t \in L_s^{\perp}, \ E\left(E_t^2\right) = 0$$

and hence

$$\forall s < t : Z_t \underset{a.s.}{=} P_{L_s}(Z_t) \in L_s.$$

Therefore,

$$\forall s \in \mathbb{Z} : Z_t \in L_s$$

and

$$Z_t \in L_{-\infty}.$$

3) "$\Leftarrow$":

$$\forall t \in \mathbb{Z} : Z_t \in L_{-\infty} \Rightarrow \forall t \in \mathbb{Z} : Z_{t+1} \in L_{-\infty} \subseteq L_t$$

and hence

$$\forall t \in \mathbb{Z} : E\left[(Z_{t+1} - P_{L_t}(Z_{t+1}))^2\right] = 0$$

so that Z_t is deterministic w.r.t. X_t.

Example 2.8 Let

$$X_t = A\cos\left(\frac{2\pi}{12}t + \varphi\right) \ (t \in \mathbb{Z}),$$

$$\varphi \sim U[-\pi, \pi] \text{ (uniform distribution on } [-\pi, \pi]),$$

$$A \text{ independent of } \varphi,\ E(A) = 0, \sigma_A^2 = var(A) < \infty.$$

Then X_t is weakly stationary and deterministic.

Example 2.9 Let

$$X_t = \varphi X_{t-12} + \varepsilon_t \ (t \in \mathbb{Z}),$$

$$\varepsilon_t \text{ iid}, E(\varepsilon_t) = 0, \sigma_\varepsilon^2 = var(\varepsilon_t) < \infty,\ |\varphi| < 1.$$

Then

$$E(X_{t+12} \mid X_s, s \le t) = \varphi X_t \in L_t$$

so that

$$\sigma_{12,\mathrm{opt}}^2 = E\left[(X_t - \varphi X_{t-12})^2\right] = \sigma_\varepsilon^2 > 0$$

and X_t is not deterministic.

Theorem 2.3 (Wold Decomposition) *Assumption:*

$$X_t \in \mathbb{R}\ (t \in \mathbb{Z})\ \textit{weakly stationary}.$$

Then

$$\exists a_0, a_1, a_2, \ldots \text{ s.t. } a_0 = 1,\ \sum_{j=0}^{\infty} a_j^2 < \infty,$$

and

$$\exists \varepsilon_t, \mu_t\ (t \in \mathbb{Z}) \text{ s.t. } \forall s, t \in \mathbb{Z}:$$

$$\varepsilon_t \in L_t,\ \mu_t \in L_{-\infty},$$

$$E(\varepsilon_t) = 0,\ cov(\varepsilon_s, \varepsilon_t) = \sigma_\varepsilon^2 \delta_{st} < \infty,\ cov(\varepsilon_s, \mu_t) = 0,$$

$$X_t \underset{a.s.,L^2(\Omega)}{=} \mu_t + \sum_{j=0}^{\infty} a_j \varepsilon_{t-j}\ (t \in \mathbb{Z}). \tag{2.1}$$

Proof

1)

$$X_t \in \mathbb{R}\ (t \in \mathbb{Z}) \text{ weakly stationary}$$

$$\Downarrow$$

$$\forall t \in \mathbb{Z}: X_t = P_{L_{t-1}}(X_t) + \varepsilon_t,\ P_{L_{t-1}}(X_t) \in L_{t-1},\ \varepsilon_t \in L_{t-1}^{\perp} \cap L_t,$$

$$\sigma_\varepsilon^2 = var(\varepsilon_t) = \sigma_{1,\text{opt}}^2,\ cov(X_s, \varepsilon_t) = \langle X_s, \varepsilon_t \rangle = 0\ (s \le t-1),$$

$$\forall s, t \in \mathbb{Z}, s \neq t: cov(\varepsilon_s, \varepsilon_t) = \langle \varepsilon_s, \varepsilon_t \rangle = 0.$$

2) Define

$$a_j = 0\ (j < 0),$$

$$a_j = \frac{\langle X_t, \varepsilon_{t-j} \rangle}{\sigma_\varepsilon^2}\ (j \ge 0).$$

In particular,

$$a_0 = \frac{\langle P_{L_{t-1}}(X_t) + \varepsilon_t, \varepsilon_t\rangle}{\sigma_\varepsilon^2} = \frac{\langle \varepsilon_t, \varepsilon_t\rangle}{\sigma_\varepsilon^2} = 1$$

and

$$X_t \text{ weakly stationary} \Rightarrow a_j \text{ independent of } t.$$

3) Define

$$E_t^0 = \left\{ Y \in L_t \mid Y = \sum_{j=1}^{k} a_j \varepsilon_{t_j},\ k \in \mathbb{N},\ a_j \in \mathbb{R},\ t_j \in \mathbb{Z},\ t_j \le t \right\},$$

$$E_t = \overline{E_t^0} = \left\{ Y \in L_t \mid \exists Y_n \in E_t^0\ (n \in \mathbb{N}) \text{ s.t. } \lim_{n\to\infty} \|Y - Y_n\|_{L^2(\Omega)}^2 = 0 \right\}. \tag{2.2}$$

Then

$$\sum_{j=0}^{\infty} a_j \varepsilon_{t-j} \in E_t \subseteq L_t$$

and hence

$$\mu_t := X_t - \sum_{j=0}^{\infty} a_j \varepsilon_{t-j} \in L_t.$$

Now,

$$\forall j \ge 0 : \langle \mu_t, \varepsilon_{t-j}\rangle = \langle X_t, \varepsilon_{t-j}\rangle - a_j \langle \varepsilon_{t-j}, \varepsilon_{t-j}\rangle = \sigma_\varepsilon^2 a_j - a_j \sigma_\varepsilon^2 = 0,$$

$$\forall j < 0 : \langle X_t, \varepsilon_{t-j}\rangle = 0,\ \langle \varepsilon_t, \varepsilon_{t-j}\rangle = 0 \Rightarrow \langle \mu_t, \varepsilon_{t-j}\rangle = 0$$

and hence

$$\forall s, t \in \mathbb{Z} : \langle \mu_s, \varepsilon_t\rangle = 0.$$

Then,

$$\forall l \ge 1 : P_{L_{t-l}}(X_t) = P_{L_{t-l}}(\mu_t) + \sum_{j=l}^{\infty} a_j \varepsilon_{t-j} = \mu_t + \sum_{j=l}^{\infty} a_j \varepsilon_{t-j} \in L_{t-l}$$

and

$$\sum_{j=l}^{\infty} a_j \varepsilon_{t-j} \in L_{t-l}.$$

Hence

$$\forall l \geq 1 : \mu_t = P_{L_{t-l}}(X_t) - \sum_{j=l}^{\infty} a_j \varepsilon_{t-j} \in L_{t-l}$$

and therefore

$$\mu_t \in L_{-\infty}.$$

Remark 2.8 Representation (2.1) is called Wold decomposition of X_t.

Remark 2.9 ε_t $(t \in \mathbb{Z})$ are only uncorrelated, not necessarily independent.

Definition 2.16 If X_t has Wold decomposition (2.1) with

$$\mu_t \equiv \mu \in \mathbb{R}, \sigma_\varepsilon^2 > 0,$$

then X_t is called purely stochastic or regular.

Example 2.10 Let

$$X_t = A \cos\left(\frac{2\pi}{12}t + \varphi\right) \ (t \in \mathbb{Z}),$$

$$\varphi \sim U[-\pi, \pi],$$

$$A \text{ independent of } \varphi, E(A) = 0, \sigma_A^2 = var(A) < \infty.$$

Then X_t has the Wold decomposition

$$X_t = A \cos\left(\frac{2\pi}{12}t + \varphi\right) = \mu_t.$$

Example 2.11 Let

$$X_t = \varepsilon_t + 2\varepsilon_{t-1},$$

$$\varepsilon_t \ (t \in \mathbb{Z}) \text{ iid}, E(\varepsilon_t) = 0, var(\varepsilon_t) = 1,$$

The X_t has the Wold decomposition

$$X_t = \sum_{j=0}^{\infty} a_j \varepsilon_{t-j} = a_0 \varepsilon_t + a_1 \varepsilon_{t-1} = \varepsilon_t + 2\varepsilon_{t-1}.$$

Example 2.12 Let

$$X_t = \sum_{j=0}^{\infty} \varphi^j \varepsilon_{t-j} \ (t \in \mathbb{Z}),$$

$$\varepsilon_t \text{ iid}, E(\varepsilon_t) = 0, \sigma_\varepsilon^2 = var(\varepsilon_t) < \infty, \ |\varphi| < 1.$$

Then X_t has the Wold decomposition

$$X_t = \sum_{j=0}^{\infty} a_j \varepsilon_{t-j}, \ a_j = \varphi^j.$$

X_t is called an autoregressive process of order 1, or $AR(1)$ process.

Example 2.13 Let

$$X_t = \sigma_t \zeta_t \ (t \in \mathbb{Z}),$$

$$\sigma_t^2 = \psi_o \xi_t^2 + \psi_1 \xi_{t-1}^2,$$

$$\zeta_t \ (t \in \mathbb{Z}) \text{ iid } N(0,1), \ \xi_t \ (t \in \mathbb{Z}) \text{ iid } N(0,1),$$

$$(\zeta_t)_{t \in \mathbb{Z}} \text{ independent of } (\xi_t)_{t \in \mathbb{Z}},$$

$$0 < \psi_o, \psi_1 < \infty.$$

Then

$$E(X_t) = 0, \ \gamma_X(k) = \delta_{0k} \cdot (\psi_o + \psi_1)$$

so that X_t has the Wold decomposition

$$X_t = \varepsilon_t = \sigma_t \zeta_t.$$

X_t is also called a Stochastic Volatility Model (SV).

Lemma 2.12 *Assumptions:*

$$X_t \in \mathbb{R} \ (t \in \mathbb{Z}) \textit{ weakly stationary, acf } \gamma_X,$$

$$X_t = \sum_{j=0}^{\infty} a_j \varepsilon_{t-j} = \textit{Wold decomposition}$$

Then

$$\gamma_X(k) = \sigma_\varepsilon^2 \sum_{j=0}^{\infty} a_j a_{j+k} \ (k \in \mathbb{Z}).$$

Proof

$$\gamma_X(k) = E\left[\left(\sum_{j=0}^{\infty} a_j \varepsilon_{t-j}\right)\left(\sum_{j=0}^{\infty} a_j \varepsilon_{t+k-j}\right)\right]$$

$$= \sum_{j,l=0}^{\infty} a_j a_l E\left(\varepsilon_{t-j}\varepsilon_{t+k-l}\right) = \sigma_\varepsilon^2 \sum_{j=0}^{\infty} a_j a_{j+k}$$

Definition 2.17 Let

$$a = \left(a_j\right)_{j\in\mathbb{N}} \in \mathbb{R}^{\mathbb{N}}, 0 < \beta < \infty.$$

Then

$$\|a\|_{\ell^\beta} = \left(\sum_{j\in\mathbb{N}} \left|a_j\right|^\beta\right)^{1/\beta},$$

$$\ell^\beta = \left\{a \mid a \in \mathbb{R}^{\mathbb{N}}, \|a\|_{\ell^\beta} < \infty\right\}.$$

For proofs of the following three Lemmas see e.g. Albiac and Kalton (2016).

Lemma 2.13

$$\|\cdot\|_{\ell^\beta} = \textit{norm} \Leftrightarrow \beta \geq 1 \Rightarrow \left(\ell^\beta, \|\cdot\|_{\ell^\beta}\right) \textit{ complete, i.e. Banach space}$$

Lemma 2.14

$$\left(\ell^\beta, \|\cdot\|_{\ell^\beta}\right) = \textit{Hilbert space} \Leftrightarrow \beta = 2 \Rightarrow \langle a, b\rangle = \sum a_j b_j$$

Lemma 2.15

$$0 < \beta < 1 \Rightarrow d(a,b) = \sum_{j\in\mathbb{N}} \left|a_j - b_j\right|^\beta = \textit{metric}$$

Notation 3

$$|a| := \left(|a_j|\right)_{j\in\mathbb{N}}$$

$$a^{(N)} := (a_0, \ldots, a_N, 0, 0, \ldots) \ (N \in \mathbb{N})$$

Lemma 2.16 *Assumptions:*

$$\varepsilon_t \in \mathbb{R} \ (t \in \mathbb{Z}), \ E(\varepsilon_t) = 0, \ cov(\varepsilon_s, \varepsilon_t) = \delta_{st}\sigma_\varepsilon^2 < \infty,$$

$$a, b \in \ell^2.$$

Then

$$X_t = \mu + \sum_{j=0}^{\infty} a_j \varepsilon_{t-j}, \ Y_t = \mu + \sum_{j=0}^{\infty} b_j \varepsilon_{t-j} \ (t \in \mathbb{Z}) \tag{2.3}$$

are weakly stationary, and

$$\|X_t - Y_t\|_{L^2(\Omega)}^2 = \sigma_\varepsilon^2 \|a - b\|_{\ell^2}^2 . \tag{2.4}$$

Proof Weak stationarity is obvious, and (2.4) follows from

$$\|X_t - Y_t\|_{L^2(\Omega)}^2 = E\left[\left(\sum_{j=0}^{\infty} (a_j - b_j)\,\varepsilon_{t-j}\right)^2\right]$$

$$= \sigma_\varepsilon^2 \sum_{j=0}^{\infty} (a_j - b_j)^2 = \sigma_\varepsilon^2 \|a - b\|_{\ell^2}^2 .$$

Lemma 2.17 *Assumptions:*

$$\varepsilon_t \in \mathbb{R} \ (t \in \mathbb{Z}), \ E(\varepsilon_t) = 0, \ cov(\varepsilon_s, \varepsilon_t) = \delta_{st}\sigma_\varepsilon^2 < \infty,$$

$$\forall n \in \mathbb{N} : a_n = \left(a_{j,n}\right)_{j\in\mathbb{N}} \in \ell^2,$$

$$X_{t,n} = \sum_{j=0}^{\infty} a_{j,n}\varepsilon_{t-j} \ (t \in \mathbb{Z}, \ n \in \mathbb{N}).$$

Then

$$X_{t,n} \ (n \in \mathbb{N}) \text{ Cauchy sequence in } L^2(\Omega) \Leftrightarrow a_n \text{ Cauchy sequence in } \ell^2$$

Proof

$$\|X_{t,n} - X_{t,m}\|_{L^2(\Omega)}^2 = E\left[\left(\sum_{j=0}^{\infty}\left(a_{j,n} - a_{j,m}\right)\varepsilon_{t-j}\right)^2\right]$$

$$= \sigma_\varepsilon^2 \sum_{j=0}^{\infty}\left(a_{j,n} - a_{j,m}\right)^2 = \sigma_\varepsilon^2 \|a_n - a_m\|_{\ell^2}^2$$

2.1.5 Ergodic Processes

Definition 2.18

$$\mathscr{F}_t \; (t \in \mathbb{Z}) = \text{sequence of } \sigma\text{-algebras on } \Omega,$$

$$\mathscr{F}_{>t} = \sigma\left(\bigcup_{s=t+1}^{\infty} \mathscr{F}_s\right) = \text{“future events”},$$

$$\mathscr{F}^{\infty} = \bigcap_{t=-\infty}^{\infty} \mathscr{F}_{>t} = \text{“asymptotic events”}.$$

Definition 2.19 Let X_t $(t \in \mathbb{Z})$ be $\mathscr{F}_t$-measurable. Then

$$X_t = \text{ergodic process}$$

$$\Leftrightarrow$$

$$\forall B \in \mathscr{F}^{\infty} : P^2(B) = P(B)$$

$$\Leftrightarrow$$

$$\forall B \in \mathscr{F}^{\infty} : P(B) \in \{0, 1\}$$

Example 2.14

$$X_t = U \; (t \in \mathbb{Z}),$$

$$U \in \mathbb{R} \text{ s.t. } 0 < p = P(U = 1) = 1 - P(U = 0) < 1,$$

$$B = \left\{ \lim_{t \to \infty} \bar{x}_n = 1 \right\} \in \mathscr{F}^{\infty} \text{ but } P(B) = p \notin \{0, 1\}.$$

$$\Rightarrow X_t \text{ not an ergodic process}$$

Theorem 2.4

$$X_t\ (t \in \mathbb{Z})\ iid \Rightarrow X_t \ ergodic\ process$$

Proof

$$X_t\ (t \in \mathbb{Z}) \text{ iid}, \mathscr{F}_t = \sigma(X_t)$$

implies

$$\forall t \in \mathbb{Z} : \mathscr{F}_t \text{ independent of } \mathscr{F}_{>t}.$$

Therefore,

$$\forall t \in \mathbb{Z} : \mathscr{F}_t \text{ independent of } \mathscr{F}^{\infty} = \bigcap_{s=-\infty}^{\infty} \mathscr{F}_{>s},$$

and

$$\sigma\left(\bigcup_{t=-\infty}^{\infty} \mathscr{F}_t\right) \text{ independent of } \mathscr{F}^{\infty}.$$

However,

$$\mathscr{F}^{\infty} \subseteq \sigma\left(\bigcup_{t=-\infty}^{\infty} \mathscr{F}_t\right) \text{ and } \mathscr{F}^{\infty} \text{ independent of } \sigma\left(\bigcup_{t=-\infty}^{\infty} \mathscr{F}_t\right)$$

implies

$$\forall B \in \mathscr{F}^{\infty} : P(B) = P(B \cap B) = P^2(B).$$

Remark 2.10 Theorem 2.4 is called Kolmogorov's $0-1$-law.

Example 2.15

$$X_t\ (t \in \mathbb{Z}) \text{ iid } N\left(0, \sigma^2\right),$$

$$B = \left\{ \lim_{n \to \infty} \bar{x}_n = 0 \right\} \in \mathscr{F}^{\infty}, P(B) = 1.$$

Example 2.16

$$X_t\ (t \in \mathbb{Z}) \text{ iid Cauchy distributed,}$$

$$B = \left\{ \lim_{n\to\infty} \bar{x}_n = 0 \right\} \in \mathscr{F}^{\infty}, P(B) = P(X_1 = 0) = 0.$$

2.1.6 *Sufficient Conditions for the a.s. Ergodic Property with a Constant Limit*

Theorem 2.5 *Assumptions:*

$$X_t\ (t \in \mathbb{Z}) \text{ strictly stationary, ergodic process, } E(|X_t|) < \infty.$$

Then

$$\mu = E(X_t) \in \mathbb{R} \text{ and } \bar{x} \underset{a.s.}{\longrightarrow} \mu$$

Proof See e.g. Walters (1982) (for the original proof see Birkhoff 1931).

Remark 2.11 Theorem 2.5 is also known as Birkhoff's ergodic theorem (in its simple version).

Problem 2.4 How does one prove that a process is ergodic?

Solution 2.4 Simple sufficient condition based on iid ε_t.

Theorem 2.6 *Assumptions:*

$$\varepsilon_t\ (t \in \mathbb{Z}) \text{ iid,}$$

$$A : \mathbb{R}^{\mathbb{N}} \to \mathbb{R} \text{ measurable,}$$

$$\forall t \in \mathbb{Z} : X_t \underset{a.s.}{=} A(\varepsilon_s, s \le t).$$

Then X_t $(t \in \mathbb{Z})$ is an ergodic process.

Proof See e.g. Stout (1974), proof of Theorem 3.5.8.

2.1.7 Sufficient Conditions for the L^2-Ergodic Property with a Constant Limit

The following conditions will be used:

(A1)

$$\forall t \in \mathbb{Z} : \mu_t = E(X_t) \in \mathbb{R}$$

(A2)

$$\forall t \in \mathbb{Z} : \sigma_t^2 = var(X_t) < \infty$$

(A3)

$$\exists \mu \in \mathbb{R} : \lim_{t \to \infty} \mu_t = \mu$$

(A4)

$$\lim_{n \to \infty} cov(\bar{x}_n, X_n) = 0$$

Definition 2.20 $a_i \in \mathbb{R}$ $(i \in \mathbb{N})$ is called Cesaro summable with limit $a \in \mathbb{R}$, if

$$\lim_{n \to \infty} \bar{a}_n = \lim_{n \to \infty} n^{-1} \sum_{i=1}^{n} a_i = a$$

Lemma 2.18 *Assumption:*

$$\exists a \in \mathbb{R} \text{ s.t. } \lim_{i \to \infty} a_i = a.$$

Then

$$\lim_{n \to \infty} \bar{a}_n = a.$$

Proof

$$\forall \varepsilon > 0 \ \exists m = m(\varepsilon) \in \mathbb{N} \text{ s.t. } \forall i > m : |a_i - a| \leq \frac{1}{2}\varepsilon.$$

Let

$$S_m = \sum_{i=1}^{m} |a_i - a|,$$

$$n > \max\left\{m(\varepsilon), 2\varepsilon^{-1} S_{m(\varepsilon)}\right\}.$$

Then

$$n^{-1}S_{m(\varepsilon)} < \frac{\varepsilon}{2}$$

and

$$\left| n^{-1}\sum_{i=1}^{n} a_i - a \right| \le n^{-1}\sum_{i=1}^{n} |a_i - a| \le n^{-1}S_{m(\varepsilon)} + \frac{n-m}{n}\frac{1}{2}\varepsilon < \varepsilon.$$

Hence

$$\lim_{n\to\infty} n^{-1}\sum_{i=1}^{n} a_i = a.$$

Theorem 2.7

$$\textit{(A1)-(A4)} \Rightarrow \lim_{n\to\infty} \|\bar{x}_n - \mu\|^2_{L^2(\Omega)} = 0.$$

Proof

1)

$$\|\bar{x}_n - \mu\|^2_{L^2(\Omega)} = B_n^2 + V_n,$$

$$B_n = n^{-1}\sum_{t=1}^{n} [\mu(t) - \mu], V_n = var(\bar{x}_n).$$

2)

$$\text{(A1)} \Rightarrow \mu(t) \text{ Cesaro summable with limit } \mu \Rightarrow \lim_{n\to\infty} B_n^2 = 0$$

3)

$$\begin{aligned} V_n &= \frac{1}{n^2}\sum_{s,t=1}^{n} cov(X_s, X_t) \\ &= \frac{2}{n^2}\sum_{t=1}^{n}\sum_{s=1}^{t} cov(X_t, X_s) - \frac{1}{n^2}\sum_{t=1}^{n} var(X_t) \\ &\le \frac{2}{n^2}\sum_{t=1}^{n}\sum_{s=1}^{t} cov(X_t, X_s) = \frac{2}{n^2}\sum_{t=1}^{n} t\, cov(\bar{x}_t, X_t) \end{aligned}$$

$$\leq \frac{2}{n}\sum_{t=1}^{n}|cov\,(\bar{x}_t, X_t)|$$

$$\text{(A4)} \;\Rightarrow cov\,(\bar{x}_t, X_t) \text{ Cesaro summable with limit 0}$$

$$\Rightarrow \lim_{n\to\infty} V_n \leq \lim_{n\to\infty} \frac{2}{n}\sum_{t=1}^{n}|cov\,(\bar{x}_t, X_t)| = 0$$

Corollary 2.5 *Assumptions:*

$$X_t\ (t \in \mathbb{Z}) \text{ weakly stationary, } E\,(X_t) = \mu, \text{ acf } \gamma_X,$$

$$\lim_{k\to\infty} \gamma_X\,(k) = 0.$$

Then

$$\lim_{n\to\infty} \|\bar{x}_n - \mu\|^2_{L^2(\Omega)} = 0.$$

Proof

1)

$$\text{weak stationarity} \implies \text{(A1), (A2), (A3)}$$

2)

$$\lim_{k\to\infty} \gamma_X\,(k) = 0 \Rightarrow \gamma_X\,(k) \text{ Cesaro summable with limit 0}$$

$$\Rightarrow \lim_{n\to\infty} cov\,(\bar{x}_n, X_n) = \lim_{n\to\infty} n^{-1}\sum_{k=0}^{n-1}\gamma_X\,(k) = 0 \Rightarrow \text{(A4)}$$

Example 2.17

$$\varepsilon_t \text{ iid, } E(\varepsilon_t) = 0,\, \sigma_\varepsilon^2 = var(\varepsilon_t) < \infty,$$

$$X_t = \beta t + \varepsilon_t$$

$$\Rightarrow \text{(A3) does not hold}$$

Example 2.18

$$\varepsilon_t \text{ iid}, E(\varepsilon_t) = 0, \sigma_\varepsilon^2 = var(\varepsilon_t) < \infty,$$

$$X_s = 0 \ (s \leq 0), X_t = \sum_{s=1}^{t} \varepsilon_s \ (t \geq 1)$$

$$\Rightarrow \text{(A4) does not hold}$$

Example 2.19

$$\varepsilon_t \text{ iid}, E(\varepsilon_t) = 0, \sigma_\varepsilon^2 = var(\varepsilon_t) < \infty,$$

$$X_t = \beta t^{-1} + \varepsilon_t$$

$$\Rightarrow \text{(A1),(A2),(A3),(A4) hold}$$

Example 2.20

$$X_t = U \ (t \in \mathbb{Z}),$$

$$U \in \mathbb{R} \text{ s.t. } 0 < p = P(U = 1) = 1 - P(U = 0) < 1,$$

$$\Rightarrow \gamma_X(k) = p(1-p) \nrightarrow 0.$$

Example 2.21

$$|\varphi| < 1, \varepsilon_t \text{ iid}, E(\varepsilon_t) = 0, \sigma_\varepsilon^2 = var(\varepsilon_t) < \infty,$$

$$X_t = \sum_{j=0}^{\infty} \varphi^j \varepsilon_{t-j}$$

$$\Rightarrow \gamma_X(k) = \frac{\varphi^{|k|}}{1-\varphi^2} \longrightarrow 0 \ (k \to \infty).$$

Remark 2.12 The condition $\lim_{k\to\infty} \gamma_X(k) = 0$ is sufficient but not necessary.

Example 2.22

$$\varphi \sim U[-\pi, \pi],$$

$$A \text{ independent of } \varphi,\ E(A) = 0,\ \sigma_A^2 = var(A) < \infty,$$

$$X_t = A(\omega)\cos\left(\frac{2\pi}{12}t + \varphi\right) \quad (t \in \mathbb{Z})$$

$$\Rightarrow \gamma_X(k) = \frac{1}{2}\sigma_A^2 \cos\left(\frac{2\pi}{12}k\right) \nrightarrow 0.$$

Nevertheless

$$\lim_{n\to\infty} \bar{x}_n = \lim_{n\to\infty} \frac{1}{n}\sum_{t=1}^{n} A\cos\left(\frac{2\pi}{12}t + \varphi\right) = 0,$$

where convergence is even pathwise.

2.2 Specific Assumptions

Here we discuss additional assumptions that are often convenient, but not as fundamental as those in the previous section.

2.2.1 Gaussian Processes

Notation 4

$$\Sigma \geq 0 \Leftrightarrow \Sigma \text{ positive semidefinite}$$

$$\Sigma > 0 \Leftrightarrow \Sigma \text{ positive definite}$$

Let $X \in \mathbb{R}^m$ be an m-dimensional Gaussian random variable with expected value μ and covariance matrix Σ. Then we write

$$X \sim N(\mu, \Sigma).$$

Remark 2.13

$$\Sigma = \text{covariance matrix} \Leftrightarrow \Sigma \text{ positive semidefinite}$$

Definition 2.21 $X_t \in \mathbb{R}^m$ $(t \in \mathbb{Z})$ is called a Gaussian time series or Gaussian process, if

$$\forall n \in \mathbb{N}, \forall t_1, \ldots, t_n \in \mathbb{Z}: \exists \mu_{t_1 \ldots t_n} \in \mathbb{R}^m, \Sigma_{t_1 \ldots t_n} \geq 0 \text{ s.t.:}$$

$$(X_{t_1}, \ldots, X_{t_m})^T \sim N(\mu_{t_1 \ldots t_m}, \Sigma_{t_1 \ldots t_m})$$

2.2.2 Linear Processes in $L^2(\Omega)$

Notation 5

$$X_{t,N} = \mu + \sum_{j=-N}^{N} a_j \varepsilon_{t-j} \; (t \in \mathbb{Z}, N \in \mathbb{N})$$

Definition 2.22 The backshift operator B is defined by

$$B : \mathbb{R}^{\mathbb{Z}} \to \mathbb{R}^{\mathbb{Z}}, x = (x_t)_{t \in \mathbb{Z}} \to y = Bx \text{ with } y_t = x_{t-1}$$

Definition 2.23 $X_t \in \mathbb{R}$ $(t \in \mathbb{Z})$ with $E(X_t) = \mu \in \mathbb{R}$ is called a linear process in $L^2(\Omega)$, if

$$\exists \varepsilon_t \in \mathbb{R} \; (t \in \mathbb{Z}) \text{ iid}, E(\varepsilon_t) = 0, \sigma_\varepsilon^2 = var(\varepsilon_t) < \infty,$$

and

$$\exists a_j \in \mathbb{R} \; (j \in \mathbb{Z}), \; \sum_{j=-\infty}^{\infty} a_j^2 < \infty$$

such that

$$\forall t \in \mathbb{Z} : \lim_{N \to \infty} \| X_t - X_{t,N} \|_{L^2(\Omega)}^2 = 0.$$

We then write

$$X_t = \mu + \sum_{j=-\infty}^{\infty} a_j \varepsilon_{t-j} = \mu + \left(\sum_{j=-\infty}^{\infty} a_j B^j \right) \varepsilon_t \; (t \in \mathbb{Z}). \tag{2.5}$$

If $a_j = 0$ $(j < 0)$, then X_t is called a causal linear process.

Lemma 2.19

$$X_t = \textit{linear process in } L^2(\Omega)$$

$$\Rightarrow X_t \textit{ weakly and strictly stationary}$$

Proof Follows directly from the definition.

Lemma 2.20

$$X_t = \textit{causal linear process in } L^2(\Omega)$$

$$\Rightarrow X_t = \mu + \sum_{j=0}^{\infty} a_j \varepsilon_{t-j} = \textit{Wold decomposition}$$

Proof Follows directly from the definition.

Corollary 2.6

$$X_t = \textit{linear process in } L^2(\Omega)$$

$$\Rightarrow \gamma_X(k) = \sigma_\varepsilon^2 \sum_{j=-\infty}^{\infty} a_j a_{j+k} \; (k \in \mathbb{Z})$$

Proof Follows as in Lemma 2.12.

Problem 2.5

$$X_t \text{ linear process in } L^2(\Omega) \underset{?}{\Rightarrow} P\left(\lim_{N\to\infty} X_{t,N} = X_t\right) = 1$$

Solution 2.5 Doob's martingale convergence theorem.

Definition 2.24 Let $Y_n \in \mathbb{R}$ s.t. $E(|Y_t|) < \infty$ $(n \in \mathbb{N})$. Then

$$Y_n = \text{supermartingale} \Leftrightarrow E(Y_n \mid Y_1, \ldots, Y_{n-1}) \le Y_{n-1} \; (n \in \mathbb{N}),$$

$$Y_n = \text{submartingale} \Leftrightarrow E(Y_n \mid Y_1, \ldots, Y_{n-1}) \ge Y_{n-1} \; (n \in \mathbb{N}),$$

$$Y_n = \text{martingale} \Leftrightarrow E(Y_n \mid Y_1, \ldots, Y_{n-1}) = Y_{n-1} \; (n \in \mathbb{N}),$$

$$Y_n = \text{martingale difference} \Leftrightarrow E(Y_n \mid Y_1, \ldots, Y_{n-1}) = 0 \; (n \in \mathbb{N}).$$

Theorem 2.8

$$Y_n \; (n \in \mathbb{N}) \textit{ supermartingale, } \sup_{n\in\mathbb{N}} E(|Y_n|) < \infty$$

$$\Rightarrow \exists Y \textit{ s.t. } E(|Y|) < \infty \textit{ and } Y_n \underset{a.s.}{\rightarrow} Y$$

Proof See e.g. Breiman (1992), Ch. 5 (also see Doob 1953).

Remark 2.14 Theorem 2.8 is called Doob's martingale convergence theorem.

Corollary 2.7

$$X_t \ (t \in \mathbb{Z}) \text{ linear process in } L^2(\Omega) \Rightarrow \forall t \in \mathbb{Z} : X_{t,N} \underset{a.s.}{\rightarrow} X_t$$

Proof

$$Y_N := X_{t,N} = \sum_{j=-N}^{N} a_j \varepsilon_{t-j} \ (N \in \mathbb{N})$$

Independence of ε_t $(t \in \mathbb{Z})$ implies

$$E(Y_N | Y_0, Y_1, \ldots, Y_{N-1}) = Y_{N-1} \Rightarrow Y_N \text{ martingale.}$$

The Cauchy-Schwarz inequality implies

$$\sup_{N\in\mathbb{N}} E(|Y_N|) \le \sup_{N\in\mathbb{N}} \sqrt{E\left(Y_N^2\right)} \le \|a\|_{\ell^2} < \infty.$$

Theorem 2.8 then implies

$$Y_N \underset{a.s.}{\rightarrow} \sum_{j=0}^{\infty} a_j \varepsilon_{t-j} = X_t.$$

Corollary 2.8

$$X_t \text{ causal linear process in } L^2(\Omega) \Rightarrow X_t \text{ ergodic process}$$

Proof

$$\text{Corollary 2.7} \Rightarrow \forall t \in \mathbb{Z} : X_t \underset{a.s.}{=} \mu + \sum_{j=0}^{\infty} a_j \varepsilon_{t-j}$$

$$\text{Theorem 2.6} \Rightarrow X_t \text{ ergodic process}$$

Corollary 2.9

$$X_t = \mu + \sum_{j=0}^{\infty} a_j \varepsilon_{t-j} \text{ linear process in } L^2(\Omega)$$

$$\Rightarrow \bar{x} \underset{a.s.}{\rightarrow} \mu$$

Proof

$$X_t \text{ strictly stationary, ergodic process, } E(|X_t|) \le \sqrt{E(X_t^2)} < \infty$$

$$\text{Birkhoff's ergodic theorem} \Rightarrow \bar{x} \underset{a.s.}{\rightarrow} \mu.$$

Lemma 2.21 *Assumptions:*

$$X_t \ (t \in \mathbb{Z}) \text{ weakly stationary}$$

$$\text{Wold decomposition } X_t = \mu + \sum_{j=0}^{\infty} a_j \varepsilon_{t-j}$$

Then

$$\lim_{k \to \infty} \gamma_X(k) = 0$$

Proof Let $k \ge 0$. Then

$$\begin{aligned} |\gamma_X(k)| &= \sigma_\varepsilon^2 \left| \sum_{j=0}^{\infty} a_j a_{j+k} \right| \le \sigma_\varepsilon^2 \sum_{j=0}^{\infty} |a_j| \, |a_{j+k}| \\ &\le \sigma_\varepsilon^2 \sqrt{\sum_{j=0}^{\infty} a_j^2 \sum_{j=0}^{\infty} a_{j+k}^2} = \sigma_\varepsilon^2 \, \|a\|_{\ell^2} \sqrt{\sum_{j=k}^{\infty} a_j^2}, \end{aligned}$$

and

$$\|a\|_{\ell^2} < \infty \Rightarrow \lim_{k \to \infty} \sum_{j=k}^{\infty} a_j^2 = 0.$$

Corollary 2.10 *Assumptions:*

$$X_t \ (t \in \mathbb{Z}) \text{ weakly stationary,}$$

$$\text{Wold decomposition } X_t = \mu + \sum_{j=0}^{\infty} a_j \varepsilon_{t-j}$$

Then

$$\bar{x} \underset{L^2(\Omega)}{\rightarrow} \mu$$

Proof

$$\lim_{k\to\infty} \gamma_X(k) = 0 \Rightarrow \text{apply Corollary 2.5}$$

Lemma 2.22

$$X_t\ (t \in \mathbb{Z})\ \textit{weakly stationary Gaussian process}$$

$$E(X_t) = \mu$$

$$\Rightarrow X_t\ (t \in \mathbb{Z})\ \textit{linear process}$$

Proof W.l.o.g. $\mu = 0$. Then the Wold decomposition is

$$X_t = \sum_{j=0}^{\infty} a_j \varepsilon_{t-j},\ \varepsilon_t = X_t - P_{L_{t-1}}(X_t),$$

and, since X_t $(t \in \mathbb{Z})$ is a Gaussian series,

$$\forall m \in \mathbb{N}, \forall t_1, \ldots, t_m \in \mathbb{Z} : (\varepsilon_{t_1}, \ldots, \varepsilon_{t_m})' \sim N\left(0, \sigma_\varepsilon^2 I\right)$$

$$\Rightarrow \varepsilon_t\ (t \in \mathbb{Z})\ \text{iid}\ N\left(0, \sigma_\varepsilon^2\right)$$

$$\Rightarrow X_t\ (t \in \mathbb{Z})\ \text{linear process}$$

2.2.3 *Linear Processes with* $E(X_t^2) = \infty$

Definition 2.25 $X_t \in \mathbb{R}$ $(t \in \mathbb{Z})$ is called a linear process in the a.s. sense, if

$$\exists \varepsilon_t \in \mathbb{R}\ (t \in \mathbb{Z})\ \text{iid},\ \exists a_j \in \mathbb{R}\ (j \in \mathbb{N})\ \text{s.t.:}$$

$$\forall t \in \mathbb{Z} : P\left(X_t = \lim_{N\to\infty} X_{t,N}\right) = 1.$$

We then write

$$X_t \underset{a.s.}{=} \sum_{j=-\infty}^{\infty} a_j \varepsilon_{t-j}\ (t \in \mathbb{Z}).$$

If $a_j = 0$ $(j < 0)$, then X_t is called a causal linear process.

Notation 6 *Let* $x_0 \in [-\infty, \infty]$. *Then*

$$f(x) \sim g(x)\ (x \to x_0) \Leftrightarrow \lim_{x \to x_0} \frac{f(x)}{g(x)} = 1$$

Also,

$$|\mu|_{\varepsilon,\beta} := E\left[|\varepsilon_t|^{\beta}\right]\ (0 < \beta < \infty)$$

Lemma 2.23

$$0 \le \beta \le 1 \Rightarrow |x+y|^{\beta} \le |x|^{\beta} + |y|^{\beta}\ (x, y \in \mathbb{R})$$

Proof

$$r := |x| + |y|$$

$$r = 0 \Rightarrow |x+y|^{\beta} = 0 = |x|^{\beta} + |y|^{\beta}$$

$$r > 0 \Rightarrow u := \frac{|x|}{r},\ v := \frac{|y|}{r}$$

$$\Rightarrow |x+y|^{\beta} \le (|x| + |y|)^{\beta} = r^{\beta}(u+v)^{\beta},\ |x|^{\beta} + |y|^{\beta} = r^{\beta}\left(u^{\beta} + v^{\beta}\right)$$

$$0 \le u, v \le 1, u + v = 1, 0 \le \beta \le 1 \Rightarrow (u+v)^{\beta} = 1 = u + v \le u^{\beta} + v^{\beta}$$

$$\Rightarrow |x+y|^{\beta} \le r^{\beta}(u+v)^{\beta} \le r^{\beta}\left(u^{\beta} + v^{\beta}\right) = |x|^{\beta} + |y|^{\beta}$$

Theorem 2.9 *Assumptions:*

$$\varepsilon_t \in \mathbb{R}\ (t \in \mathbb{Z})\ iid,\ \varepsilon_t \sim F_{\varepsilon},\ \exists f_{\varepsilon} = F'_{\varepsilon},$$

$$\exists \alpha \in (0,2),\ 0 < C_1, C_2 < \infty\ s.t.\ 1 - F_{\varepsilon}(x) \sim C_1 x^{-\alpha},\ F_{\varepsilon}(-x) \sim C_2 x^{-\alpha}\ (x \to \infty)$$

$$a = \left(a_j\right)_{j \in \mathbb{Z}} \in \mathbb{R}^{\mathbb{Z}}\ s.t.\ \exists \beta \in I_{\alpha} : a \in \ell^{\beta}$$

where

$$I_{\alpha} = (0, \alpha)\ (0 < \alpha \le 1)\ and\ I_{\alpha} = (0, 1]\ (1 < \alpha < 2).$$

Then

$$\exists X_t \ (t \in \mathbb{Z}) \text{ strictly stationary s.t. } \forall t \in \mathbb{Z} : P\left(\lim_{N\to\infty} X_{t,N} = X_t\right) = 1.$$

Proof

1) w.l.o.g.

$$a_j = 0 \ (j < 0).$$

2)

$$1 - F_\varepsilon(x) \sim C_1 x^{-\alpha}, \ F_\varepsilon(-x) \sim C_2 x^{-\alpha} \ (x \to \infty)$$

$$\Rightarrow f_\varepsilon(x) \sim \alpha C_1 x^{-\alpha-1}, f_\varepsilon(-x) \sim \alpha C_2 x^{-\alpha-1} \ (x \to \infty)$$

$$\Rightarrow \forall \delta \in (0, \alpha) : |\mu|_{\varepsilon,\delta} = E\left(|\varepsilon_t|^\delta\right) < \infty$$

3) $1 < \alpha < 2$: Let

$$Y_{t,N} = \sum_{j=0}^{N} |a_j| \, |\varepsilon_{t-j}| \, .$$

Then

$$0 \le |X_{t,N}| \le Y_{t,N} \le Y_{t,N+1},$$

$$E(Y_{t,N}) = |\mu|_{\varepsilon,1} \sum_{j=0}^{N} |a_j| \le |\mu|_{\varepsilon,1} \, \|a\|_{\ell^1} < \infty$$

and hence

$$\exists \lim_{N\to\infty} E(Y_{t,N}) = |\mu|_{\varepsilon,1} \, \|a\|_{\ell^1}$$

$$\Rightarrow P\left(\exists \lim_{N\to\infty} Y_{t,N} \in \mathbb{R}\right) = 1$$

$$\Rightarrow P\left(\exists \lim_{N\to\infty} X_{t,N} \in \mathbb{R}\right) = 1.$$

3) $0 < \alpha \leq 1$: Let

$$\beta \in I_\alpha = (0, \alpha) \text{ s.t. } |\mu|_{\varepsilon,\beta} < \infty,\ a \in \ell^\beta,$$

and define

$$Y_{t,N} = \sum_{j=0}^{N} \left|a_j\right|^\beta \left|\varepsilon_{t-j}\right|^\beta .$$

Then

$$0 \leq Y_{t,N} \leq Y_{t,N+1},$$

$$E\left(Y_{t,N}\right) = |\mu|_{\varepsilon,\beta} \sum_{j=0}^{N} \left|a_j\right|^\beta \leq |\mu|_{\varepsilon,\beta} \|a\|_{\ell^\beta}^\beta < \infty$$

and hence

$$\exists \lim_{N\to\infty} E\left(Y_{t,N}\right) = |\mu|_{\varepsilon,\beta} \|a\|_{\ell^\beta}^\beta$$

$$\Rightarrow P\left(\exists \lim_{N\to\infty} Y_{t,N} \in \mathbb{R}\right) = 1.$$

Moreover

$$0 < \beta < 1 \Rightarrow |X_{t,N}|^\beta = \left|\sum_{j=0}^{N} a_j \varepsilon_{t-j}\right|^\beta \leq \sum_{j=0}^{N} \left|a_j\right|^\beta \left|\varepsilon_{t-j}\right|^\beta = Y_{t,N}$$

so that

$$P\left(\exists \lim_{N\to\infty} X_{t,N} \in \mathbb{R}\right) = 1$$

Lemma 2.24

X_t ($t \in \mathbb{Z}$) causal linear process in the a.s. sense $\Rightarrow X_t$ ergodic process

Proof

$$\forall t \in \mathbb{Z} : X_t \underset{a.s.}{=} \mu + \sum_{j=0}^{\infty} a_j \varepsilon_{t-j}$$

Theorem 2.6 $\Rightarrow X_t$ ergodic process

Example 2.23

$$\varepsilon_t \; (t \in \mathbb{Z}) \text{ iid Cauchy} \Rightarrow \alpha = 1 \Rightarrow I_\alpha = (0, 1)$$

$$|a_j| = O\left(|j|^{-1-\delta}\right) \text{ for some } \delta > 0 \Rightarrow P\left(\exists \lim_{N\to\infty} X_{t,N} \in \mathbb{R}\right) = 1$$

Corollary 2.11 *Assumptions:*

$$X_t = \mu + \sum_{j=0}^{\infty} a_j \varepsilon_{t-j} \; (t \in \mathbb{Z}) \text{ causal linear process in the a.s. sense}$$

$$E(|X_t|) < \infty$$

Then

$$E(X_t) = \mu \in \mathbb{R} \text{ and } P\left(\exists \lim_{n\to\infty} \bar{x} = \mu\right) = 1$$

Proof

$$X_t \; (t \in \mathbb{Z}) \text{ a.s. causal linear process} \Rightarrow X_t \text{ ergodic process}$$

$$\text{Birkhoff's ergodic theorem} \Rightarrow \bar{x} \underset{a.s.}{\rightarrow} \mu$$

2.2.4 Multivariate Linear Processes

Notation 7

$$M(m, l, \mathbb{R}) = \text{set of real valued } m \times l \text{ matrices } (l, m \in \mathbb{N})$$

$$GL(m, \mathbb{R}) = \text{set of invertible real valued } m \times m \text{ matrices } (m \in \mathbb{N})$$

$$\|x - y\|^2_{\mathbb{R}^m} = \sqrt{\sum_{i=1}^{m} (x_i - y_i)^2} \; (x, y \in \mathbb{R}^m)$$

Definition 2.26 Let

$$X = (X_1, \ldots, X_m)', Y = (Y_1, \ldots, Y_m)' \; (X_i, Y_i \in L^2(\Omega)).$$

Then

$$\|X - Y\|_{L^2(\Omega)}^2 = E\left[\|X(\omega) - Y(\omega)\|_{\mathbb{R}^m}^2\right]$$

Definition 2.27 $X_t \in \mathbb{R}^m$ $(t \in \mathbb{Z})$ with $E(X_t) = \mu \in \mathbb{R}^m$ is called a linear process in $L^2(\Omega)$, if

$$\exists \varepsilon_t \in \mathbb{R}^l \ (t \in \mathbb{Z}) \text{ iid s.t.:}$$

$$E(\varepsilon_t) = 0, \ \Sigma_\varepsilon = var(\varepsilon_t) = E\left(\varepsilon_t \varepsilon_t'\right) \in GL(l, \mathbb{R}),$$

and

$$\exists A_j \in M(m, l, \mathbb{R}) \ \ (j \in \mathbb{Z})$$

such that

$$\forall t \in \mathbb{Z} : \lim_{N \to \infty} \|X_t - X_{t,N}\|_{L^2(\Omega)}^2 = 0,$$

where

$$X_{t,N} = \mu + \sum_{j=-N}^{N} A_j \varepsilon_{t-j}.$$

We then write

$$X_t \underset{L^2(\Omega)}{=} \mu + \sum_{j=-\infty}^{\infty} A_j \varepsilon_{t-j} \ (t \in \mathbb{Z}).$$

If $A_j = 0$ $(j < 0)$, then X_t is called a causal linear process.

Remark 2.15 The definition of m-dimensional linear processes in the a.s. sense is analogous.

2.2.5 Invertibility

Recall the definitions

$$L_t^0 = \left\{ Y \mid Y = \sum_{j=1}^{n} a_j X_{t_j}, n \in \mathbb{N}, a_j \in \mathbb{R}, t_j \in \mathbb{Z}, t_j \le t \right\},$$

$$L_t = \overline{L_t^0}, L_{-\infty} = \cap_{t=-\infty}^{\infty} L_t,$$

$$E_t^0 = \left\{ Y \in L_t \mid Y = \sum_{j=1}^{n} a_j \varepsilon_{t_j}, n \in \mathbb{N}, a_j \in \mathbb{R}, t_j \in \mathbb{Z}, t_j \leq t \right\},$$

$$E_t = \overline{E_t^0},$$

and Lemma 2.9 with

$$\sigma_k^2 \left(\hat{X}_{t+k} \right) = \sigma_{k,\text{opt}}^2 \Leftrightarrow \hat{X}_{t+k} = P_{L_t} \left(X_{t+k} \right).$$

Problem 2.6 How can $P_{L_t}(X_{t+k})$ and $\sigma_{k,\text{opt}}^2$ be computed, given X_s $(s \leq t)$? In particular, is $\hat{X}_{t+k}$ of the form

$$\hat{X}_{t+k} = \mu_t + \sum_{j=0}^{\infty} \beta_j (k) X_{t-j}? \tag{2.6}$$

Solution 2.6 General results on orthogonal projections in Hilbert spaces.

A first partial answer is given by the following Lemma. Since μ_t can be predicted perfectly, we will assume $\mu_t \equiv 0$ w.l.o.g.

Lemma 2.25 *Assumptions:*

$$X_t \ (t \in \mathbb{Z}) \ \textit{weakly stationary}$$

$$X_t = \sum_{j=0}^{\infty} a_j \varepsilon_{t-j} = \textit{Wold decomposition.}$$

Then

$$P_{L_t} \left(X_{t+k} \right) = \sum_{j=0}^{\infty} a_{j+k} \varepsilon_{t-j} \tag{2.7}$$

and

$$\sigma_{k,opt}^2 = \sigma_\varepsilon^2 \sum_{j=0}^{k-1} a_j^2.$$

Proof

$$\varepsilon_t \in L_t, X_t = \sum_{j=0}^{\infty} a_j \varepsilon_{t-j} \in E_t \ (t \in \mathbb{Z})$$

implies

$$E_t = L_t$$

and hence

$$P_{L_t}(X_{t+k}) = P_{E_t}(X_{t+k}).$$

Since

$$\begin{aligned} X_{t+k} &= \sum_{j=0}^{\infty} a_j \varepsilon_{t+k-j} = \sum_{j=-k}^{\infty} a_{j+k} \varepsilon_{t-j} \\ &= A_t + B_t \end{aligned}$$

with

$$A_t = \sum_{j=0}^{\infty} a_{j+k} \varepsilon_{t-j} \in E_t, \; B_t = \sum_{j=-k}^{-1} a_{j+k} \varepsilon_{t-j} \in E_t^{\perp},$$

we obtain

$$P_{L_t}(X_{t+k}) = P_{E_t}(X_{t+k}) = A_t.$$

Moreover,

$$\sigma_{k,\text{opt}}^2 = E\left(B_t^2\right) = \sigma_\varepsilon^2 \sum_{j=0}^{k-1} a_j^2.$$

Remark 2.16

$$\lim_{k\to\infty} c_{k,\text{opt}}^2 = \sigma_\varepsilon^2 \sum_{j=0}^{\infty} a_j^2 = \gamma_X(0)$$

Problem 2.7

$$\sum_{j=0}^{\infty} a_{j+k} \varepsilon_{t-j} \underset{?}{=} \sum_{j=0}^{\infty} \beta_j(k) X_{t-j}$$

Solution 2.7 In infinite dimensional Hilbert spaces this turns out to be true under orthogonality assumptions.

Lemma 2.26 *Assumptions:*

$$H = \textit{Hilbert space}, \ v_j \in H \ (j = 1, \ldots, n)$$

$$V_n^0 = \left\{ x \in H \mid x = \sum_{j=1}^{n} a_j v_j, \ a_j \in \mathbb{R} \right\} = span\left(v_j, 1 \le j \le n\right)$$

$$V_n = \overline{V_n^0}$$

Then

$$V_n = V_n^0,$$

i.e. every finite dimensional linear subspace is closed.

Proof See e.g. Rynne and Youngson (2007).

Corollary 2.12 *Assumptions: Conditions of Lemma 2.26, and*

$$u \in H.$$

Then

$$\exists b_{j,n} \in \mathbb{R} \ (j = 1, \ldots, n) \ s.t. \ P_{V_n}(u) = \sum_{j=1}^{n} b_{j,n} v_j.$$

Proof

$$P_{V_n}(u) \in V_n = V_n^0$$

together with the orthogonalization

$$e_1 = v_1, e_2 = v_2 - \frac{\langle v_2, e_1 \rangle}{\|e_1\|^2} e_1, \ldots, e_n = v_n - \sum_{j=1}^{n-1} \frac{\langle v_n, e_j \rangle}{\|e_j\|^2} e_j$$

and

$$b_{j,n} = \frac{\langle u, e_j \rangle}{\|e_j\|^2} \ (j = 1, \ldots, n)$$

imply

$$P_{V_n} = (u) \sum_{j=1}^{n} b_{j,n} v_j.$$

Conclusion 3 *For finite-dimensional subspaces V_n,*

$$P_{V_n}(u) = \sum_{j=1}^{n} b_{j,n} v_j.$$

Note however that $b_{j,n}$ may change with n.

For infinite dimensional subspaces the situation is more complicated.

Lemma 2.27 *Assumptions:*

$$H = \text{Hilbert space},\ v_j \in H\ (j \in \mathbb{N})$$

$$V^0 = \bigcup_{n \in \mathbb{N}} V_n,\ V = \overline{V^0}$$

$$u \in H,\ \hat{u}_n = P_{V_n}(u)\ (n \in \mathbb{N}),\ \hat{u} = P_V(u).$$

Then

$$\lim_{n \to \infty} \|\hat{u}_n - \hat{u}\| = 0.$$

Proof Follows directly from the definitions.

Conclusion 4 *The projection on a closed linear subspace V can be approximated by projections to the finite dimensional spaces V_n. Note however,*

$$\hat{u}_n = \sum_{j=1}^{n} b_{j,n} v_j \underset{L^2(\Omega)}{\rightarrow} \hat{u}$$

does not necessarily imply:

$$\exists b_j \in \mathbb{R}\ (1 \le j \le n) \text{ s.t. } \hat{u} = \sum_{j=1}^{\infty} b_j v_j.$$

Example 2.24

$$\varepsilon_t\ (t \in \mathbb{Z}) \text{ iid},\ E(\varepsilon_t) = 0,\ var(\varepsilon_t) = 1$$

$$u = X_t = \varepsilon_t - \varepsilon_{t-1}\ (t \in \mathbb{Z})$$

$$v_j = X_{t-j} \; (j = 1, 2, \ldots)$$

$$V_n = span\,(v_1, \ldots, v_n)$$

Then

$$H = L^2(\Omega),$$

$$E(X_t) = 0,\; \gamma_X(0) = 2,\; \gamma_X(1) = -1,\; \gamma_X(k) = 0\; (|k| \geq 2),$$

$$V^0 = \bigcup_{n \geq 1} V_n^0 = L_{t-1}^0,\; V = \overline{V^0} = L_{t-1},$$

$$\hat{u}_n = P_{V_n}(X_t),\; \hat{u} = P_V(X_t) = P_{L_{t-1}}(X_t)$$

Corollary 2.12 implies

$$\hat{u}_n = \sum_{j=1}^{n} b_{j,n} v_j$$

with

$$0 = \left\langle X_t - \sum_{j=1}^{n} b_{j,n} X_{t-j}, X_{t-l} \right\rangle = \gamma(l) - \sum_{j=1}^{n} b_{j,n} \gamma(l-j) \;\; (l = 1, \ldots, n).$$

The solution is

$$b_{j,n} = \frac{j}{n+1} - 1 \; (j = 1, \ldots, n),$$

so that

$$\hat{u}_n = \sum_{j=1}^{n} \left(\frac{j}{n+1} - 1 \right) v_j.$$

If

$$\exists b_j \in \mathbb{R} \; (j \in \mathbb{Z}) \text{ s.t. } \lim_{n \to \infty} \left\| \hat{u} - \sum_{j=1}^{n} b_j v_j \right\|^2 = 0,$$

then we would have

$$\lim_{n \to \infty} \left\| \hat{u}_n - \sum_{j=1}^{n} b_j v_j \right\|^2 = 0. \tag{2.8}$$

Now

$$\hat{u}_n - \sum_{j=1}^{n} b_j v_j = \sum_{j=1}^{n} \left(\frac{j}{n+1} - 1 - b_j\right) \left(\varepsilon_{t-j} - \varepsilon_{t-j-1}\right)$$

$$= -\left(\frac{n}{n+1} + b_1\right) \varepsilon_{t-1} + \sum_{j=1}^{n-1} \left(\frac{1}{n+1} + b_j - b_{j+1}\right) \varepsilon_{t-j-1} + \left(\frac{1}{n+1} + b_n\right) \varepsilon_{t-n-1}$$

so that

$$\left\|\hat{u}_n - \sum_{j=1}^{n} b_j v_j\right\|^2 = \left(\frac{n}{n+1} + b_1\right)^2 + \sum_{j=1}^{n-1} \left(\frac{1}{n+1} + b_j - b_{j+1}\right)^2 + \left(\frac{1}{n+1} + b_n\right)^2 . \tag{2.9}$$

Convergence of $\sum_{j=1}^{n} b_j v_j$, and hence $b_n \to 0$, together with (2.9) lead to

$$\liminf_{n} \left\|\hat{u}_n - \sum_{j=1}^{n} b_j v_j\right\|^2 \geq (1 + b_1)^2 + \sum_{j=1}^{\infty} \left(b_j - b_{j+1}\right)^2 .$$

For (2.8) we thus need

$$(1 + b_1)^2 = 0$$

and

$$\sum_{j=1}^{\infty} \left(b_j - b_{j+1}\right)^2 = 0,$$

and hence

$$b_j = -1 \ (j \in \mathbb{N}).$$

However,

$$\left\|\hat{u}_n - \sum_{j=1}^{n} (-1)\, v_j\right\|^2 = \frac{1}{(n+1)^2} \left\|\sum_{j=1}^{n} j \left(\varepsilon_{t-j} - \varepsilon_{t-j-1}\right)\right\|^2$$

$$= \frac{1}{(n+1)^2} \left\|\sum_{j=1}^{n} \varepsilon_{t-j} - n\varepsilon_{t-n-1}\right\|^2$$

$$= \frac{n + n^2}{(n+1)^2}$$

so that

$$\liminf \left\| \hat{u}_n - \sum_{j=1}^{n} (-1)\, v_j \right\|^2 = 1.$$

Hence

$$\nexists b_j \in \mathbb{R}\ (j \in \mathbb{Z}) \text{ s.t. } \lim_{n\to\infty} \left\| \hat{u} - \sum_{j=1}^{n} b_j v_j \right\|^2 = 0.$$

Remark 2.17 The problem is that in general v_j ($j \in \mathbb{N}$) are not orthogonal. For orthogonal v_j's, the problem disappears.

Lemma 2.28 *Assumptions:*

$$H = \text{Hilbert space, } V_n \text{ as above}$$

$$\langle v_j, v_l \rangle = \delta_{jl}\ (j, l \in \mathbb{N})$$

$$u \in H,\ \hat{u}_n = P_{V_n}(u)\ \ (n \in \mathbb{N})$$

$$b_j = \langle u, v_j \rangle\ (j \in \mathbb{N})$$

Then

$$\forall n \in \mathbb{N} : \hat{u}_n = \sum_{j=1}^{n} b_j v_j$$

Proof

$$\sum_{j=1}^{n} b_j v_j \in V_n,$$

$$z \in V_n \Rightarrow z = \sum_{j=1}^{n} c_j v_j$$

Hence

$$\left\langle z, u - \sum_{j=1}^{n} b_j v_j \right\rangle = \sum_{l=1}^{n} c_l \left[\langle v_l, u \rangle - \sum_{j=1}^{n} b_j \langle v_l, v_j \rangle \right]$$

$$= \sum_{l=1}^{n} c_l (b_l - b_l) = 0.$$

Thus

$$\sum_{j=1}^{n} b_j v_j = P_{V_n}(u).$$

Conclusion 5 *For orthogonal v_j, the coefficients b_j in $P_{V_n}(u)$ do not depend on n.*

Corollary 2.13 *Assumptions:*

$$H = \text{Hilbert space, } V_n, V \text{ as above}$$

$$\langle v_j, v_l \rangle = \delta_{jl} \; (j, l \in \mathbb{N})$$

$$u \in H, \; \hat{u}_n = P_{V_n}(u) \;\; (n \in \mathbb{N}), \; \hat{u} = P_V(u)$$

$$b_j = \langle u, v_j \rangle \; (j \in \mathbb{N})$$

Then

$$\hat{u} = \sum_{j=1}^{\infty} b_j v_j.$$

Proof From Lemma 2.28

$$\hat{u}_n = P_{V_n}(u) = \sum_{j=1}^{n} b_j v_j,$$

and

$$V = \overline{\bigcup_{n \in \mathbb{N}} V_n}$$

implies

$$\lim_{n \to \infty} \|\hat{u} - \hat{u}_n\| = 0,$$

and therefore,

$$\lim_{n \to \infty} \left\| \sum_{j=1}^{n} b_j v_j - \hat{u} \right\| = 0.$$

Conclusion 6

$$v_j \textit{ orthogonal} \Rightarrow P_V(u) = \sum_{j=1}^{\infty} b_j v_j.$$

Problem 2.8 In general, X_t $(t \in \mathbb{Z})$ are not orthogonal, so that

$$\sum_{j=0}^{\infty} b_{j,n} X_{t-j} \underset{L^2(\Omega)}{\rightarrow} P_{L_t}(X_{t+k})$$

may not necessarily imply

$$P_{L_t}(X_{t+k}) = \sum_{j=0}^{\infty} b_j X_{t-j}.$$

Solution 2.8 Introduce the assumption of invertibility. The definition is simple for linear processes.

Definition 2.28 Let

$$\varepsilon_t \ (t \in \mathbb{Z}) \text{ iid}, E(\varepsilon_t) = 0, \sigma_\varepsilon^2 = var(\varepsilon_t) < \infty,$$

$$\sum_{j=-\infty}^{\infty} a_j^2 < \infty, \mu \in \mathbb{R},$$

$$X_t = \mu + \sum_{j=-\infty}^{\infty} a_j \varepsilon_{t-j} \ (t \in \mathbb{Z}).$$

Then the process X_t is called invertible with respect to ε_t, if

$$\exists b_j \in \mathbb{R} \ (j \in \mathbb{N}) \text{ s.t. } \forall t \in \mathbb{Z} : \varepsilon_t \underset{L^2(\Omega), a.s.}{=} \sum_{j=0}^{\infty} b_j \left(X_{t-j} - \mu \right).$$

Remark 2.18 Invertibility is not an "absolute" property of the process. It always refers to a specific iid sequence ε_t $(t \in \mathbb{Z})$ only. Sometimes the process X_t may be invertible w.r.t. some other iid $\tilde{\varepsilon}_t$.

Example 2.25

$$\varepsilon_t \ (t \in \mathbb{Z}) \text{ iid}, E(\varepsilon_t) = 0, E\left(\varepsilon_t^2\right) = \sigma_\varepsilon^2 < \infty,$$

$$X_t = \varepsilon_{t-1} \ (t \in \mathbb{Z})$$

$$\Rightarrow \varepsilon_t = X_{t+1}$$

$$\Rightarrow X_t \text{ not invertible w.r.t. } \varepsilon_t, \text{ invertible w.r.t. } \tilde{\varepsilon}_t = \varepsilon_{t-1}$$

Example 2.26

$$\varepsilon_t \; (t \in \mathbb{Z}) \text{ iid}, \; E(\varepsilon_t) = 0, \; E\left(\varepsilon_t^2\right) = \sigma_\varepsilon^2 < \infty, \; |\varphi| < 1$$

$$X_t = \sum_{j=0}^{\infty} \varphi^j \varepsilon_{t-j}$$

$$\Rightarrow X_t = \varphi X_{t-1} + \varepsilon_t \; (t \in \mathbb{Z}) \Rightarrow \varepsilon_t = X_t - \varphi X_{t-1} \; (t \in \mathbb{Z})$$

$$\Rightarrow X_t \text{ invertible w.r.t. } \varepsilon_t \; (t \in \mathbb{Z}).$$

Example 2.27

$$\varepsilon_t \; (t \in \mathbb{Z}) \text{ iid}, \; E(\varepsilon_t) = 0, \; E\left(\varepsilon_t^2\right) = \sigma_\varepsilon^2 < \infty$$

$$X_t = \varepsilon_t - \varepsilon_{t-1}$$

$$\Rightarrow X_t \text{ not invertible w.r.t. } \varepsilon_t \; (t \in \mathbb{Z})$$

Remark 2.19 For nonlinear processes, X_t is a nonlinear function of iid ε_t's. Therefore, a more complicated definition of invertibility is needed (see e.g. Tong 1990). It is however possible to define invertible linear filters, and invertibility of the Wold decomposition.

Definition 2.29 Let

$$a_j \in \mathbb{R} \; (j \in \mathbb{Z}) \text{ s.t. } \sum_{j=-\infty}^{\infty} a_j^2 < \infty.$$

Then the operator

$$A : \mathbb{R}^{\mathbb{Z}} \to \mathbb{R}^{\mathbb{Z}}, y \to Ay = \left(\sum_{j=-\infty}^{\infty} a_j y_{t-j} \right)_{t \in \mathbb{Z}}$$

is called a linear filter in $L^2(\Omega)$. We also write

$$A = A(B) = \sum_{j=-\infty}^{\infty} a_j B^j,$$

$$Ay = A(B)\, y,$$

and, for

$$z = A(B)\, y = (z_t)_{t\in\mathbb{Z}},$$

$$z_t = A(B)\, y_t.$$

Lemma 2.29 *Assumptions:*

$$\varepsilon_t \in \mathbb{R}\ (t \in \mathbb{Z}),\ E(\varepsilon_t) = 0,\ cov(\varepsilon_s, \varepsilon_t) = \sigma_\varepsilon^2 \delta_{st} < \infty$$

$$A(B) = \sum_{j=-\infty}^{\infty} a_j B^j = \textit{linear filter in } L^2(\Omega)$$

Then

$$\forall t \in \mathbb{Z} : X_t = \sum_{j=-\infty}^{\infty} a_j \varepsilon_{t-j} \in L^2(\Omega).$$

Proof

$$a \in \ell^2 \Rightarrow X_{t,N} \text{ Cauchy sequence in } L^2(\Omega)$$

$$L^2(\Omega) \text{ complete} \Rightarrow X_{t,N} \underset{L^2(\Omega)}{\rightarrow} X_t = \sum_{j=-\infty}^{\infty} a_j \varepsilon_{t-j} \in L^2(\Omega)$$

Definition 2.30 Let A be a linear filter in $L^2(\Omega)$. Then A is called invertible, if

$$\exists A^{-1} = \sum_{j=-\infty}^{\infty} b_j B^j = \text{linear filter in } L^2(\Omega) \text{ s.t.:}$$

$$X_t = A(B)\, \varepsilon_t = \sum_{j=-\infty}^{\infty} a_j \varepsilon_{t-j}\ (t \in \mathbb{Z})$$

with

$$\varepsilon_t \in \mathbb{R}\ (t \in \mathbb{Z}), E(\varepsilon_t) = 0, cov\ (\varepsilon_s, \varepsilon_t) = \sigma_\varepsilon^2 \delta_{st} < \infty,$$

implies

$$\varepsilon_t = A^{-1}(B)\, X_t = \sum_{j=-\infty}^{\infty} b_j X_{t-j}.$$

Remark 2.20 In Definition 2.30, ϵ_t may also depend on future values of X_s $(s \in \mathbb{Z})$.

Definition 2.31 A Wold decomposition

$$X_t = A\,(B)\, \varepsilon_t$$

is called invertible, if A is an invertible filter and

$$A^{-1}(B) = \sum_{j=0}^{\infty} b_j B^j.$$

Remark 2.21 For a linear process with Wold decomposition $X_t = \sum_{j=0}^{\infty} a_j \varepsilon_{t-j}$ and iid ε_t, we have

$$X_t \text{ invertible process} \Leftrightarrow \text{Wold decomposition invertible}$$

Corollary 2.14 *Assumptions:*

$$X_t\ (t \in \mathbb{Z}) \text{ weakly stationary}$$

$$X_t = A\,(B)\, \varepsilon_t = \text{invertible Wold decomposition}$$

$$\varepsilon_t = A^{-1}(B)\, X_t = \sum_{j=0}^{\infty} b_j X_{t-j}\ (t \in \mathbb{Z})$$

$$\beta_j\,(k) = \sum_{l=0}^{j} a_{l+k} b_{j-l}\ (k \in \mathbb{N})$$

$$\hat{X}_{t+k} = \sum_{j=0}^{\infty} \beta_j\,(k)\, X_{t-j}$$

Then

$$E\left[\left(X_{t+k}-\hat{X}_{t+k}\right)^{2}\right]=\sigma_{k,opt}^{2}=\inf_{Y\in L_{t}}\sigma_{k}^{2}\left(Y\right).$$

Proof

1) For

$$\hat{X}_{t+k}=\sum_{l=0}^{\infty}a_{l+k}\varepsilon_{t-l},$$

Lemma 2.25 implies

$$E\left[\left(X_{t+k}-\hat{X}_{t+k}\right)^{2}\right]=\inf_{Y\in E_{t}}E\left[\left(X_{t+k}-Y\right)^{2}\right]=\inf_{Y\in L_{t}}E\left[\left(X_{t+k}-Y\right)^{2}\right]$$

2) Invertibility of the Wold decomposition implies

$$\varepsilon_{t-l}=\sum_{r=0}^{\infty}b_{r}X_{t-l-r}\in L_{t-l}$$

and hence

$$\hat{X}_{t+k}=\sum_{l=0}^{\infty}a_{l+k}\left(\sum_{r=0}^{\infty}b_{r}X_{t-l-r}\right)=\sum_{j=0}^{\infty}\beta_{j}\left(k\right)X_{t-j}.$$

with

$$\beta_{j}\left(k\right)=\sum_{l=0}^{j}a_{l+k}b_{j-l}.$$

Example 2.28

$$\varepsilon_{t}\ (t\in\mathbb{Z})\text{ iid},\ E\left(\varepsilon_{t}\right)=0,\ E\left(\varepsilon_{t}^{2}\right)=\sigma_{\varepsilon}^{2}<\infty,\ -1<\varphi<1$$

$$X_{t}=\sum_{j=0}^{\infty}\varphi^{j}\varepsilon_{t-j},\ a_{j}=\varphi^{j}$$

$$\Rightarrow\varepsilon_{t}=X_{t}-\varphi X_{t-1}$$

$$\Rightarrow b_{0}=1,\ b_{1}=-\varphi,\ b_{l}=0\ (l\geq2)$$

$$\Rightarrow \beta_j(k) = \sum_{l=0}^{j} a_{l+k} b_{j-l} = \sum_{l=0}^{j} \varphi^{l+k} b_{j-l} = \varphi^k \left(-\varphi^{j-1} \cdot \varphi + \varphi^j \cdot 1\right) = 0 \; (j \geq 1),$$

$$\beta_0(k) = \varphi^k$$

$$\Rightarrow \hat{X}_{t+k} = \varphi^k X_t.$$

2.2.6 Restrictions on the Dependence Structure

There are many possibilities how to impose general restrictions on the dependence structure. A few examples are discussed in the following.

2.2.6.1 Markov Processes

Definition 2.32 Let $X_t \in E$ ($t \in \mathbb{Z}$) be measurable w.r.t. $\mathscr{F}_t$. Then X_t is called a Markov process with state space E, if

$$\forall A \in \mathscr{F}_t, \forall k \in \mathbb{N}, \forall t_1 < \ldots < t_k < t \in \mathbb{Z}:$$

$$P(X_t \in A \mid X_{t_1}, \ldots, X_{t_k}) \underset{a.s.}{=} P(X_t \in A \mid X_{t_k})$$

If E is finite or countable, then X_t is called a Markov chain.

Lemma 2.30

$$X_t \in \mathbb{R}^m \; (t \in \mathbb{Z}) \textit{ Markov process}$$

$$\Rightarrow \forall k \geq 1: E(X_{t+k} \mid X_s, s \leq t) \underset{a.s.}{=} E(X_{t+k} \mid X_t)$$

Proof Follows from the definition.

Remark 2.22

$$X_t \; (t \in \mathbb{Z}) \text{ Markov process} \not\Rightarrow \forall k \geq 2: X_t, X_{t-k} \text{ independent}$$

Remark 2.23

$$\forall j \geq 2: X_t, X_{t-j} \text{ independent} \not\Rightarrow X_t \; (t \in \mathbb{Z}) \text{ Markov process}$$

Example 2.29

$$\varepsilon_t \ (t \in \mathbb{Z}) \text{ iid}, E(\varepsilon_t) = 0, E\left(\varepsilon_t^2\right) = \sigma_\varepsilon^2 < \infty, \ -1 < \varphi < 1$$

$$X_t = \sum_{j=0}^{\infty} \varphi^j \varepsilon_{t-j}$$

$$\Rightarrow X_t = \varphi X_{t-1} + \varepsilon_t$$

$$\Rightarrow X_t = \text{Markov process}, cov(X_t, X_{t-2}) = \varphi^2 var(X_{t-2}) \neq 0$$

Example 2.30

$$\varepsilon_t \ (t \in \mathbb{Z}) \text{ iid } N\left(0, \sigma_\varepsilon^2\right), \ -1 < \psi < 1$$

$$X_t = \psi \varepsilon_{t-1} + \varepsilon_t$$

Then

$$\forall j \geq 2 : X_t, X_{t-j} \text{ independent.}$$

Moreover,

$$X_t = \text{Gaussian process}, E(X_t) = 0$$

and

$$\rho = corr(X_t, X_{t-1}) = \frac{\psi}{1 + \psi^2}$$

implies

$$E(X_t \mid X_{t-1}) = \rho X_{t-1}.$$

However,

$$X_t = \sum_{j=0}^{\infty} (-1)^{j+1} \psi^j X_{t-j} + \epsilon_t$$

which implies

$$E(X_t \mid X_s, s \leq t-1) = \sum_{j=0}^{\infty} (-1)^{j+1} \psi^j X_{t-j}.$$

Hence, for $\psi \neq 0$,

$$E(X_t \mid X_s, s \leq t-1) \neq E(X_t \mid X_{t-1})$$

and

$$X_t \neq \text{Markov process.}$$

Remark 2.24 Some processes become Markov processes when written in state space representation. Typical examples are discussed below.

Example 2.31 Let

$$\varepsilon_t\ (t \in \mathbb{Z}) \text{ iid}, E(\varepsilon_t) = 0, \sigma_\varepsilon^2 = var(\varepsilon_t) < \infty,$$

$$\varphi(z) = 1 - \sum_{j=1}^{p} \varphi_j z^j\ (z \in \mathbb{C}) \text{ s.t. } \varphi(z) \neq 0\ (|z| \leq 1).$$

In Chap. 6, we show:

$$\exists X_t\ (t \in \mathbb{Z}) \text{ weakly and strictly stationary s.t.:}$$

$$\varphi(B) X_t = \varepsilon_t\ (t \in \mathbb{Z}).$$

Then

$$p \geq 2, \max_{2 \leq j \leq p} |\varphi_j| > 0 \Rightarrow X_t \neq \text{Markov process}$$

However,

$$Y_t = AY_{t-1} + u_{t-1} \in \mathbb{R}^p\ (t \in \mathbb{Z})$$

with

$$Y_t = \begin{pmatrix} X_{t-p+1} \\ X_{t-p+2} \\ \vdots \\ X_t \end{pmatrix}, u_t = \begin{pmatrix} 0 \\ 0 \\ \vdots \\ 0 \\ \varepsilon_t \end{pmatrix},$$

$$\underset{p\times p}{A}=\begin{pmatrix} 0 & 1 & 0 & \cdots & 0 \\ \vdots & 0 & 1 & \ddots & \vdots \\ \vdots & \vdots & \ddots & \ddots & 0 \\ 0 & 0 & \cdots & 0 & 1 \\ \varphi_p & \varphi_{p-1} & \cdots & \cdots & \varphi_1 \end{pmatrix}.$$

Hence,

$$Y_t = \text{Markov process.}$$

Remark 2.25

$$\text{Markov process} \not\Leftrightarrow \text{martingale}$$

Example 2.32

$$\varepsilon_t\ (t \in \mathbb{Z})\ \text{iid},\ E(\varepsilon_t) = 0,\ E\left(\varepsilon_t^2\right) = \sigma_\varepsilon^2 < \infty,\ -1 < \varphi < 1$$

$$X_t = \sum_{j=0}^{\infty} \varphi^j \varepsilon_{t-j}$$

$$\Rightarrow X_t = \text{Markov process},\ E(X_t \mid X_s, s \le t-1) = \varphi X_{t-1} \neq X_{t-1}$$

$$\Rightarrow X_t \neq \text{martingale}$$

Example 2.33

$$\varepsilon_t\ (t \in \mathbb{Z})\ \text{iid},\ E(\varepsilon_t) = 0,\ E\left(\varepsilon_t^2\right) = \sigma_\varepsilon^2 < \infty,\ -1 < \varphi < 1$$

$$X_t = X_{t-1} + \varepsilon_t\ (t \in \mathbb{N})$$

$$\Rightarrow X_t = \text{Markov process and martingale}$$

Example 2.34

$$\varepsilon_t\ (t \in \mathbb{Z})\ \text{iid},\ E(\varepsilon_t) = 0,\ E\left(\varepsilon_t^2\right) = \sigma_\varepsilon^2 < \infty,$$

$$X_0 = X_1 = 0$$

$$X_t = X_{t-1} + (1 + X_{t-2})\,\varepsilon_t\ (t \in \mathbb{N}).$$

Then

$$E\left(X_t \mid X_s, s \leq t-1\right) = X_{t-1}$$

so that

$$X_t = \text{martingale.}$$

However

$$X_t \neq \text{Markov process.}$$

Example 2.35

$$\varepsilon_t \ (t \in \mathbb{N}) \text{ iid}, E(\varepsilon_t) = 0, \sigma_\varepsilon^2 = var(\varepsilon_t) < \infty$$

$$a, b_j > 0 \ (j \geq 1), \sigma_\varepsilon^2 \sum_{j=1}^{\infty} b_j < 1$$

$$\Rightarrow \exists X_t \text{ strictly stationary s.t.:}$$

$$X_t = \left(a + \sum_{j=1}^{\infty} b_j X_{t-j}\right) \varepsilon_t \ (t \in \mathbb{Z})$$

(see ARCH(∞) processes in Chap. 7). We then have

$$E\left(X_t \mid X_s, s \leq t-1\right) = 0 \Rightarrow X_t = \text{martingale difference}$$

and

$$\max_{j \geq 2} \left|b_j\right| > 0 \Rightarrow X_t \neq \text{Markov process}$$

Example 2.36

$$\varepsilon_t \ (t \in \mathbb{Z}) \text{ iid } N\left(0, \sigma_\varepsilon^2\right), \psi \neq 0$$

$$X_t = \psi \varepsilon_{t-1} + \varepsilon_t \ (t \in \mathbb{Z}).$$

$$\Rightarrow X_t \neq \text{Markov process}, X_t \neq \text{martingale}$$

2.2.6.2 Mixing Conditions

Definition 2.33 Given a probability space $(\Omega, \mathscr{F}, P)$ and σ-algebras$\mathscr{A}, \mathscr{B} \subseteq \mathscr{F}$, define

$$\alpha_0(\mathscr{A}, \mathscr{B}) = \sup\{|P(A \cap B) - P(A)P(B)| : A \in \mathscr{A}, B \in \mathscr{B}\}.$$

Definition 2.34 Let

$$X_t \ (t \in \mathbb{Z}) \ \mathscr{F}_t\text{-measurable},$$

$$\mathscr{F}_{\leq t} = \sigma\left(\bigcup_{s=-\infty}^{t} \mathscr{F}_s\right), \mathscr{F}_{\geq t} = \sigma\left(\bigcup_{s=t}^{\infty} \mathscr{F}_s\right).$$

Then X_t is called strongly mixing, if

$$\alpha(n) = \sup_{t \in \mathbb{Z}} \alpha_0(\mathscr{F}_{\leq t}, \mathscr{F}_{\geq t+n}) \underset{n\to\infty}{\to} 0.$$

Remark 2.26 Definition 2.34 was introduced in Rosenblatt (1956). Many other mixing conditions exist. Mixing conditions are useful for proving central limit theorems under general conditions (see e.g. Doukhan 1995, Bradley 2007, McKean 2014).

2.2.6.3 Short Memory, Long Memory, Antipersistence

Definition 2.35 A weakly stationary process X_t $(t \in \mathbb{Z})$ with acf γ_X is called a

$$\text{short-memory process} \Leftrightarrow 0 < \sum_{k=-\infty}^{\infty} \gamma_X(k) < \infty,$$

$$\text{long-memory process} \Leftrightarrow \sum_{k=-\infty}^{\infty} \gamma_X(k) = \infty,$$

$$\text{antipersistent process} \Leftrightarrow \sum_{k=-\infty}^{\infty} \gamma_X(k) = 0$$

(see e.g. Beran 1994, Beran et al. 2013, Giraitis et al. 2012).

Remark 2.27 More refined definitions exist that also include the notion of intermediate dependence (see Definition 6.7 in Chap. 6).

Chapter 3
Defining Probability Measures for Time Series

Recall:

$$\text{Time series model} = (\Omega, \mathscr{F}, P)$$

$$\Omega = \left(\mathbb{R}^k\right)^T = \text{space of functions } X : T \to \mathbb{R}^k \ (k \in \mathbb{N}, T \subseteq \mathbb{R})$$

$$P = \text{probability distribution on } (\Omega, \mathscr{F})$$

Problem 3.1 Suppose we wish that P satisfies certain conditions. Does such a probability measure P on $(\Omega, \mathscr{F})$ exist? Typical conditions are for instance specific finite dimensional distributions, equations, transformations, stationarity, invertibility, causality.

Solution 3.1 Kolmogorov's fundamental consistency result, and other, more constructive, methods for defining time series.

3.1 Finite Dimensional Distributions

Notation 8

$$\mathbf{x} = (x_1, \ldots, x_m)^T \in \mathbb{R}^m, \mathbf{t} = (t_1, \ldots, t_m)^T \in \mathbb{Z}^m$$

$$\mathbf{x}_{-i} = \left(x_j\right)_{j=1,\ldots,m; j \neq i}, \mathbf{t}_{-i} = \left(t_j\right)_{j=1,\ldots,m; j \neq i}$$

$$\mathbf{X_t} = (X_{t_1}, \ldots, X_{t_m})^T, \mathbf{X}_{\mathbf{t}_{-i}} = \left(X_{t_j}\right)_{j=1,\ldots,m; j \neq i}$$

J. Beran, *Mathematical Foundations of Time Series Analysis*,
https://doi.org/10.1007/978-3-319-74380-6_3

$$F_{\mathbf{t}}(\mathbf{x}) = P(\mathbf{X}_{\mathbf{t}} \le \mathbf{x}) = P(X_{t_1} \le x_1, \ldots, X_{t_m} \le x_m)$$

$$F_{\mathbf{t}_{-i}}(\mathbf{x}) = P(\mathbf{X}_{\mathbf{t}_{-i}} \le \mathbf{x}) \ (\mathbf{x} \in \mathbb{R}^{m-1})$$

$$\mathbf{u} = (u_1, \ldots, u_m)^T \in \mathbb{R}^m$$

$$\varphi_{\mathbf{t}}(\mathbf{u}) = E\left[\exp\left(i\mathbf{u}^T\mathbf{X}_{\mathbf{t}}\right)\right], \ \varphi_{\mathbf{t}_{-i}}(\mathbf{u}_{-i}) = E\left[\exp\left(i\mathbf{u}_{-i}^T\mathbf{X}_{\mathbf{t}_{-i}}\right)\right]$$

Theorem 3.1 *Assumption:*

$$\mathscr{P}_{finite} = \{F_{\mathbf{t}} \mid \mathbf{t} \in \mathbb{Z}^m, \ m \in \mathbb{N}\}$$

Then

$$\exists P \text{ on } \mathbb{R}^{\mathbb{Z}} \text{ with } \mathscr{P}_{finite}$$

$$\Leftrightarrow \forall \mathbf{x} \in \mathbb{R}^m, \mathbf{t} \in \mathbb{Z}^m, i \in \{1, \ldots, m\} : \lim_{x_i \to \infty} F_{\mathbf{t}}(\mathbf{x}) = F_{\mathbf{t}_{-i}}(\mathbf{x})$$

$$\Leftrightarrow \forall \mathbf{u} \in \mathbb{R}^m, \mathbf{t} \in \mathbb{Z}^m, i \in \{1, \ldots, m\} : \lim_{u_i \to 0} \varphi_{\mathbf{t}}(\mathbf{u}) = \varphi_{\mathbf{t}_{-i}}(\mathbf{u}_{-i})$$

Proof See e.g. Bhattacharya and Waymire (2007). For the original theorem see Kolmogorov (1956).

Remark 3.1 Theorem 3.1 is known as Kolmogorov's consistency theorem or Kolmogorov's extension theorem. It can be generalized to P on more general spaces, such as $\mathbb{R}^{\mathbb{R}}$ or $\mathbb{R}^{[0,T]}$.

3.2 Transformations and Equations

Transformations of known processes provide a flexible tool for defining new time series models. The proof of existence essentially reduces to the proof of measurability of the transformation, and possibly the existence of some moments. Typical examples are:

- linear processes as transformations of iid ε_t
- random walk type processes with

$$X_t = X_t + \varepsilon_t \ (t \in \mathbb{N})$$

- transformations in the spectral domain (see Chap. 4), i.e. given

$$X_t = \int_{-\pi}^{\pi} e^{it\lambda} dZ_X(\lambda)$$

define

$$Y_t = \int_{-\pi}^{\pi} g(\lambda) e^{it\lambda} dZ_X(\lambda).$$

Equations are often motivated by specific applications. The existence of solutions with desired properties is not guaranteed and has to be proved.

Example 3.1 Let

$$\varepsilon_t \text{ iid}, E(\varepsilon_t) = 0, \sigma_\varepsilon^2 = var(\varepsilon_t) < \infty, \ -1 < \varphi < 1.$$

Then

$$X_t = \varphi X_{t-1} + \varepsilon_t \ (t \in \mathbb{Z})$$

has the weakly and strictly stationary solution

$$X_t = \sum_{j=0}^{\infty} \varphi^j \varepsilon_{t-j}.$$

Example 3.2 Let

$$\varepsilon_t \in \mathbb{R} \ (t \in \mathbb{Z}) \text{ iid}, \sigma_\varepsilon^2 = E\left(\varepsilon_t^2\right) = 1,$$

$$p, q \in \mathbb{N}, \alpha = \left(\alpha_0, \alpha_1, \ldots, \alpha_p\right)^T \in \mathbb{R}_+, \alpha_0 > 0, \beta = \left(\beta_1, \ldots, \beta_q\right)^T \in \mathbb{R}_+,$$

$$\sigma_t^2 = \alpha_0 + \sum_{j=1}^{p} \alpha_j X_{t-j}^2 + \sum_{j=1}^{q} \beta_j \sigma_{t-j}^2 \ (t \in \mathbb{Z}).$$

Then a strictly stationary solution of

$$X_t = \sigma_t \varepsilon_t \ (t \in \mathbb{Z})$$

exists under conditions given in Theorem 7.8 (see GARCH(p, q) processes in Chap. 7).

3.3 Conditions on the Expected Value

Problem 3.2

$$\mu : \mathbb{Z} \to \mathbb{R}, t \to \mu(t)$$

$$\underset{?}{\Rightarrow} \exists X_t \ (t \in \mathbb{Z} \text{ or } \mathbb{N}) \text{ s.t. } E(X_t) = \mu(t)$$

Solution 3.2 No conditions needed in discrete time.

Lemma 3.1

$$\mu : \mathbb{Z} \to \mathbb{R}$$

$$\Rightarrow \exists X_t \ (t \in \mathbb{Z}) \ s.t. \ \forall t \in \mathbb{Z} : E(X_t) = \mu(t)$$

Proof For instance $X_t = \mu(t)$ $(t \in \mathbb{Z})$.

Problem 3.3

$$X_t \in \mathbb{R} \ (t \in \mathbb{Z}), \ E(X_t) = \mu(t)$$

$$\Downarrow ?$$

$$\exists \hat{\mu}_n : \mathbb{R}^n \to \mathbb{R}^{\mathbb{Z}}, \ (x_1, \ldots, x_n) \to \hat{\mu}_n(\cdot; x_1, \ldots, x_n) \text{ s.t.:}$$

$$\forall t \in \mathbb{Z} : \hat{\mu}_n(t; X_1, \ldots, X_n) \underset{p}{\to} \mu(t)$$

Solution 3.3 Depends on $\mu(t)$ and properties of $\varepsilon_t = X_t - \mu(t)$.

Example 3.3

$$0 < p = P(U = -1) = 1 - P(U = 1) < 1$$

$$X_t = \mu + \varepsilon_t = \mu + U \ (t \in \mathbb{Z})$$

$$\Rightarrow \nexists \hat{\mu}_n \text{ s.t. } \hat{\mu}_n \underset{p}{\to} \mu$$

Example 3.4

$$\varepsilon_t \ (t \in \mathbb{Z}) \text{ iid}, \ E(\varepsilon_t) = 0, \ var(\varepsilon_t) < \infty$$

$$X_t = \mu + \varepsilon_t \ (t \in \mathbb{Z})$$

$$\Rightarrow \hat{\mu}_n = \bar{x} = n^{-1} \sum_{t=1}^{n} X_t \underset{p}{\to} \mu$$

Example 3.5

$$\mu(t) = \beta t^{-\alpha} \ (\alpha > 0, \beta \in \mathbb{R}, t \geq 1)$$

$$\varepsilon_t \ (t \in \mathbb{N}) \text{ iid } N(0, 1)$$

$$X_t = \mu(t) + \varepsilon_t \ (t \in \mathbb{N})$$

Then the minimum variance unbiased estimator (MVUE) of β is

$$\hat{\beta} = \frac{\sum_{t=1}^{n} X_t t^{-\alpha}}{\sum_{t=1}^{n} t^{-2\alpha}}$$

with

$$E\left(\hat{\beta}\right) = \beta,\ var\left(\hat{\beta}\right) = \frac{1}{\sum_{t=1}^{n} t^{-2\alpha}}.$$

Therefore,

$$\exists \hat{\mu}_n \text{ s.t. } \forall t \in \mathbb{Z} : \hat{\mu}_n(t) \underset{p}{\rightarrow} \mu(t) \Leftrightarrow \sum_{t=1}^{\infty} t^{-2\alpha} = \infty \Leftrightarrow \alpha \leq \frac{1}{2}.$$

3.4 Conditions on the Autocovariance Function

Problem 3.4

$$\gamma : \mathbb{Z} \times \mathbb{Z} \rightarrow \mathbb{R},\ (s, t) \rightarrow \gamma(s, t)$$

$$\underset{?}{\Rightarrow} \exists X_t\ (t \in \mathbb{Z}) \text{ s.t. } cov(X_s, X_t) = \gamma(s, t)$$

Solution 3.4 Necessary and sufficient conditions can be derived as outlined in the following.

3.4.1 *Positive Semidefinite Functions*

Definition 3.1

$$\gamma : \mathbb{Z} \times \mathbb{Z} \rightarrow \mathbb{R}$$

is called a symmetric function, if

$$\forall s, t \in \mathbb{Z} : \gamma(t, s) = \gamma(s, t).$$

Similarly,

$$\gamma : \mathbb{Z} \times \mathbb{Z} \rightarrow \mathbb{C}$$

is called a Hermitian function, if

$$\forall s, t \in \mathbb{Z} : \gamma(t, s) = \overline{\gamma(s, t)}.$$

Also,

$$\gamma : \mathbb{Z} \to \mathbb{R}$$

is called a symmetric function, if

$$\forall k \in \mathbb{Z} : \gamma(-k) = \gamma(k),$$

and

$$\gamma : \mathbb{Z} \to \mathbb{C}$$

is called a Hermitian function, if

$$\forall k \in \mathbb{Z} : \gamma(-k) = \overline{\gamma(k)}.$$

Definition 3.2

$$\gamma : \mathbb{Z} \times \mathbb{Z} \to \mathbb{R}$$

is called a positive semidefinite function, if

$$\forall n \in \mathbb{N}, \mathbf{a} \in \mathbb{R}^n, \mathbf{t} \in \mathbb{Z}^n : \sum_{i,j=1}^{n} a_i a_j \gamma\left(t_i, t_j\right) \geq 0.$$

Similarly,

$$\gamma : \mathbb{Z} \times \mathbb{Z} \to \mathbb{C}$$

is called a positive semidefinite function, if

$$\forall n \in \mathbb{N}, \mathbf{a} \in \mathbb{C}^n, \mathbf{t} \in \mathbb{Z}^n : \sum_{i,j=1}^{n} a_i \overline{a_j} \gamma\left(t_i, t_j\right) \geq 0.$$

If ">" holds for all $\mathbf{a} \neq 0$, then γ is called positive definite. Also,

$$\gamma : \mathbb{Z} \to \mathbb{R}$$

is called a positive semidefinite function, if

$$\forall n \in \mathbb{N}, \mathbf{a} \in \mathbb{R}^n, \mathbf{t} \in \mathbb{Z}^n : \sum_{i,j=1}^{n} a_i a_j \gamma\left(t_i - t_j\right) \geq 0,$$

and

$$\gamma : \mathbb{Z} \to \mathbb{C}$$

is called a positive semidefinite function, if

$$\forall n \in \mathbb{N}, \mathbf{a} \in \mathbb{C}^n, \mathbf{t} \in \mathbb{Z}^n : \sum_{i,j=1}^{n} a_i \overline{a_j} \gamma\left(t_i - t_j\right) \geq 0.$$

Lemma 3.2

$$\forall n \in \mathbb{N}, \mathbf{a} \in \mathbb{R}^n, \mathbf{t} \in \mathbb{Z}^n : \sum_{i,j=1}^{n} a_i \overline{a_j} \gamma\left(t_i, t_j\right) \geq 0$$

$$\Leftrightarrow$$

$$\forall n \in \mathbb{N}, \mathbf{a} \in \mathbb{R}^n : \sum_{i,j=1}^{n} a_i \overline{a_j} \gamma\left(i, j\right) \geq 0$$

Proof "$\Rightarrow$":

$$\sum_{i,j=1}^{n} a_i \overline{a_j} \gamma\left(i, j\right) = \sum_{i,j=1}^{n} a_i \overline{a_j} \gamma\left(t_i - t_j\right) \text{ with } \mathbf{t} = (1, \ldots, n)^T$$

"$\Leftarrow$": Let

$$\mathbf{t} \in \mathbb{Z}^n, \mathbf{a} \in \mathbb{R}^n$$

Set

$$m = \max_{1 \leq j \leq n} t_j, \mathbf{b} = (b_1, \ldots, b_m)^T \in \mathbb{R}^m$$

with

$$b_{t_j} = a_j \ (j = 1, \ldots, n), b_l = 0 \text{ otherwise}.$$

Then

$$0 \leq \sum_{i,j=1}^{n} b_i \overline{b_j} \gamma(i,j) = \sum_{i,j=1}^{n} a_i \overline{a_j} \gamma(t_i, t_j)$$

Theorem 3.2

$$\exists X_t \in \mathbb{R}\ (t \in \mathbb{Z})\ s.t.\ \forall s,t \in \mathbb{Z} : cov(X_s, X_t) = \gamma(s,t)$$

$$\Leftrightarrow$$

$$\gamma : \mathbb{Z} \times \mathbb{Z} \to \mathbb{R}\ \textit{symmetric, positive semidefinite}$$

Proof

1) "$\Rightarrow$":

$$X_s, X_t \in \mathbb{R},\ \gamma_X(s,t) = cov(X_s, X_t) \Rightarrow \gamma_X(s,t) = \gamma_X(t,s)$$

$$n \in \mathbb{N}, \mathbf{a} \in \mathbb{R}^n,\ Y = \sum_{i=1}^{n} a_i X_i \Rightarrow 0 \leq var(Y) = \sum_{i,j=1}^{n} a_i a_j \gamma_X(i,j)$$

2) "$\Leftarrow$": Define

$$\Sigma_{\mathbf{t}} := \left[\gamma\left(t_j, t_l\right)\right]_{j,l=1,\ldots,n}\ (n \in \mathbb{N}, \mathbf{a} \in \mathbb{R}^n, \mathbf{t} \in \mathbb{Z}^n).$$

Then

$$\gamma \text{ symmetric, positive semidefinite} \Rightarrow \Sigma_{\mathbf{t}} = \text{covariance matrix.}$$

Hence

$$\exists \mathbf{Y_t} = \left(Y_{t_1} \cdots Y_{t_n}\right)^T \sim N(0, \Sigma_{\mathbf{t}}).$$

Defining

$$F_{\mathbf{t}}(\mathbf{y}) = P(\mathbf{Y_t} \leq \mathbf{y})\ (n \in \mathbb{N}, \mathbf{t} \in \mathbb{Z}^n, \mathbf{y} \in \mathbb{R}^n),$$

$$\mathscr{P}_{\text{finite}} = \{F_{\mathbf{t}} \mid n \in \mathbb{N}, \mathbf{t} \in \mathbb{Z}^n\},$$

the result follows from Theorem 3.1.

Theorem 3.3

$$\exists X_t \in \mathbb{C}\ (t \in \mathbb{Z})\ s.t.\ \forall s,t \in \mathbb{Z} : cov\,(X_s, X_t) = \gamma\,(s,t)$$

$$\Leftrightarrow$$

$$\gamma : \mathbb{Z} \times \mathbb{Z} \to \mathbb{C}\ \textit{Hermitian, positive semidefinite}$$

Corollary 3.1

$$\exists X_t \in \mathbb{R}\ (t \in \mathbb{Z})\ s.t.\ \forall k \in \mathbb{Z} : cov\,(X_t, X_{t+k}) = \gamma\,(k)$$

$$\Leftrightarrow$$

$$\gamma : \mathbb{Z} \to \mathbb{R}\ \textit{symmetric positive semidefinite}$$

3.4.2 Spectral Distribution

Problem 3.5 Simple conditions for positive semidefiniteness?

Solution 3.5 Spectral representation of γ.

Example 3.6

$$\gamma\,(0) \geq 0,\ \gamma\,(k) = 0\ (k \neq 0)$$

$$\Rightarrow \sum_{i,j=1}^{n} a_i a_j \gamma\,(i-j) = \gamma\,(0) \sum_{i=1}^{n} a_i^2 \geq 0$$

$$\Rightarrow \gamma \text{ is an acf}$$

Example 3.7

$$\gamma\,(0) = \sigma^2 \geq 0,\ \frac{\gamma\,(\pm 1)}{\gamma\,(0)} = \frac{\psi}{1+\psi^2}\ (\psi \in \mathbb{R}),\ \gamma(k) = 0\ (\,|k| \geq 2)$$

$$\Rightarrow \sum_{i,j=1}^{n} a_i a_j \gamma(i-j) = \sigma^2 \left(\sum_{i=1}^{n} a_i^2 + 2\frac{\psi}{1+\psi^2} \sum_{i=1}^{n-1} a_i a_{i+1} \right)$$

Define

$$a = (a_1, \dots, a_n)^T, b = (a_2, \dots, a_n, 0)^T.$$

Then

$$\sum_{i,j=1}^{n} a_i a_j \gamma(i-j) = \sigma^2 \left(\|a\|^2 + 2\frac{\psi}{1+\psi^2} \langle a, b \rangle \right) \geq \sigma^2 \|a\|^2 \left(1 - \frac{2|\psi|}{1+\psi^2} \right).$$

Then

$$\forall \psi \in \mathbb{R} : \left| \frac{2\psi}{1+\psi^2} \right| \leq 1$$

implies

$$\forall \psi \in \mathbb{R} : \gamma \text{ is an acf}$$

Example 3.8 Let

$$d \in \left(-\frac{1}{2}, \frac{1}{2} \right),$$

$$\gamma(k) = \frac{\Gamma(|k| + d)\Gamma(1-d)}{\Gamma(|k| - d + 1)\Gamma(d)} \quad (k \in \mathbb{Z}).$$

Checking positive semidefiniteness of γ directly is more difficult than in the previous examples.

Conclusion 7 *In general it may not be easy to check whether a function γ is positive semidefinite. An easier way is however provided by the so-called spectral decomposition discussed in the following.*

Notation 9 *Let*

$$F : \mathbb{R} \to \mathbb{R}, A \subseteq \mathbb{R}.$$

Then we write

$$F\left(x^+\right) = \lim_{\varepsilon \downarrow 0} F(x+\varepsilon), \; F\left(x^-\right) = \lim_{\varepsilon \downarrow 0} F(x-\varepsilon), \; \Delta F(x) = F\left(x^+\right) - F\left(x^-\right),$$

$$J_{cont}(F; A) = \{x \in A | \Delta F(x) = 0\} = \{x \in A | F \text{ continuous at } x\},$$

$$J_{deriv}(F; A) = \left\{ x \in A | \; \exists f(x) = \frac{d}{dx} F(x) \right\} = \{x \in A | F \text{ differentiable at } x\},$$

$$J_{jump^+}(F;A) = \{x \in A| \; \exists \Delta F(x), \; \Delta F(x) > 0\},$$

$$J_{jump^-}(F;A) = \{x \in A| \; \exists \Delta F(x), \; \Delta F(x) < 0\},$$

$$J_{jump}(F;A) = J_{jump^+}(F;A) \cup J_{jump^-}(F;A),$$

$$J_{cont}(F) = J_{cont}(F;\mathbb{R}), \; J_{deriv}(F) = J_{deriv}(F;\mathbb{R}),$$

$$C(A) = \textit{set of continuous functions on A},$$

$$C_b(A) = \textit{set of bounded continuous functions on A}.$$

Definition 3.3 A function

$$F : \mathbb{R} \to \mathbb{R}$$

is called a generalized cumulative distribution function (gcdf) on $\mathbb{R}$, if

$$\forall x \in \mathbb{R} : \exists F\left(x^+\right) \text{ and } F(x) = F\left(x^+\right) \text{ (right-continuous)},$$

$$\forall x, y \in \mathbb{R} \text{ s.t. } x \leq y : F(x) \leq F(y) \text{ (nondecreasing)},$$

$$F(-\infty) = 0, \; F(\infty) < \infty \text{ (nonnegative, bounded)}$$

First we state some elementary results.

Lemma 3.3 *Assumption:*

$$F \textit{ gcdf on } \mathbb{R}.$$

Then

$$\forall x \in \mathbb{R} : \exists F(x^-) \in \mathbb{R}_+ \textit{ and } \Delta F(x) = F(x) - F(x^-).$$

Lemma 3.4 *Assumptions:*

$$F \textit{ gcdf on } \mathbb{R}$$

$$A_1 = J_{deriv}(F), \; A_2 = J_{cont}(F) \diagdown J_{deriv}(F), \; A_3 = J_{jump^+}(F)$$

Then

$$A_i \cap A_j = \varnothing \; (i \neq j)$$

and

$$\mathbb{R} = A_1 \cup A_2 \cup A_3.$$

Remark 3.2 The result means that, for each $x \in \mathbb{R}$, only one of the following three cases is possible:

$$(1)\ \exists f(x) = F'(x)\,;\ (2)\ \Delta F(x) = 0 \text{ and } \nexists F'(x)\,;\ (3)\ \Delta F(x) > 0.$$

Lemma 3.5 *Assumption:*

$$F \text{ gcdf on } \mathbb{R}.$$

Then

$$J_{jump+}(F) \text{ is finite or countable.}$$

Lemma 3.6 *Assumptions:*

$$F \text{ gcdf on } \mathbb{R}$$

$$g : \mathbb{R} \to \mathbb{R} \text{ bounded}$$

$$a, b \in \mathbb{R},\ a \le b$$

$$J_{jump}(g; [a, b]) \text{ finite}$$

$$J_{jump}(g; [a, b]) \cap J_{jump+}(F; [a, b]) = \varnothing$$

Then

$$\exists \text{Riemann-Stieltjes integral} \int_a^b g(\lambda)\, dF(\lambda) \in \mathbb{R}$$

Proof See e.g. Johnsonbaugh and Pfaffenberger (2010) (also see Stieltjes 1894).

Lemma 3.7 *Under the assumptions of Lemma 3.6,*

$$\int_a^b g(\lambda)\, dF(\lambda) = g(b) F(b) - g(a) F(a) - \int_a^b F(\lambda)\, dg(\lambda)$$

Proof See e.g. Johnsonbaugh and Pfaffenberger (2010).

Definition 3.4

$$L^2(F;\mathbb{R}) = \left\{ h:\mathbb{R}\to\mathbb{R} \mid \|h\|^2 = \int_{-\pi}^{\pi} |h(\lambda)|^2 \, dF(\lambda) < \infty \right\},$$

$$\langle g,h\rangle_{L^2(F;\mathbb{R})} = \int_{-\pi}^{\pi} g(\lambda)\, h(\lambda)\, dF(\lambda),$$

$$L^2(F;\mathbb{C}) = \left\{ h:\mathbb{C}\to\mathbb{C} \mid \|h\|^2 = \int_{-\pi}^{\pi} |h(\lambda)|^2 \, dF(\lambda) < \infty \right\},$$

$$\langle g,h\rangle_{L^2(F;\mathbb{C})} = \int_{-\pi}^{\pi} g(\lambda)\, \overline{h(\lambda)} dF(\lambda).$$

More generally, for $p \geq 1$,

$$L^p(F;\mathbb{R}) = \left\{ h:\mathbb{R}\to\mathbb{R} \mid \|h\|^p = \int_{-\pi}^{\pi} |h(\lambda)|^p \, dF(\lambda) < \infty \right\},$$

$$L^2(F;\mathbb{C}) = \left\{ h:\mathbb{C}\to\mathbb{C} \mid \|h\|^p = \int_{-\pi}^{\pi} |h(\lambda)|^p \, dF(\lambda) < \infty \right\},$$

If it is clear whether $\mathbb{C}$ or $\mathbb{R}$ is used, then we also write $L^p(F)$.

Lemma 3.8 *$L^2(F;\mathbb{C})$ and $L^2(F;\mathbb{R})$ are Hilbert spaces.*

Proof See e.g. Bruckner et al. (1997).

Definition 3.5 Let

$$F, F_n\ (n \in \mathbb{N}) \text{ gcdf on } \mathbb{R},\ X_n \sim F_n.$$

Then X_n converges in distribution to $X \sim F$, or F_n converges to F, if

$$\forall x \in J_{cont}(F): \lim_{n\to\infty} F_n(x) = F(x).$$

We write

$$X_n \underset{d}{\rightarrow} X \text{ or } F_n \rightarrow F.$$

Lemma 3.9 *Assumptions:*

$$F, F_n\ (n \in \mathbb{N}) \text{ gcdf on } \mathbb{R}$$

Then

$$F_n \rightarrow F \Leftrightarrow \forall g \in C_b(\mathbb{R}): \lim_{n\to\infty} \int g(x)\, dF_n(x) = \int g(x)\, dF(x)$$

Proof See e.g. Ash (1972).

Lemma 3.10 *Assumptions:*

$$F_n\ (n\in\mathbb{N})\ gcdf\ on\ \mathbb{R}$$

$$\exists M<\infty\ s.t.\ F_n(\infty)\le M\ (n\in\mathbb{N})$$

Then

$$\exists\ gcdf\ F\ and\ a\ subsequence\ F_{n_j}\ s.t.\ F_{n_j}\to F$$

Proof See e.g. Ash (1972).

Remark 3.3 Lemma 3.10 is known as Helly's selection theorem.

Theorem 3.4

$$\gamma:\mathbb{Z}\to\mathbb{C}\ positive\ semidefinite$$

$$\Leftrightarrow$$

$$\exists\ gcdf\ F\ s.t.:$$

$$F(-\pi)=0,\ F(\pi)=F(\infty),$$

$$\gamma(k)=\int_{-\pi}^{\pi}e^{ik\lambda}dF(\lambda)\ \ (k\in\mathbb{Z})$$

Proof

1) "$\Leftarrow$": Let

$$\gamma(k)=\int_{-\pi}^{\pi}e^{ik\lambda}dF(\lambda).$$

Then

$$\gamma(-k)=\int_{-\pi}^{\pi}e^{-ik\lambda}dF(\lambda)=\overline{\int_{-\pi}^{\pi}e^{ik\lambda}dF(\lambda)}=\overline{\gamma(k)}$$

$$\Rightarrow\gamma\ \text{Hermitian.}$$

Moreover,

$$\sum_{j,l=1}^{n} a_j \overline{a_l} \gamma (j-l) = \int_{-\pi}^{\pi} \sum_{j,l=1}^{n} a_j \overline{a_l} e^{i(j-l)\lambda} dF(\lambda)$$

$$= \int_{-\pi}^{\pi} \left| \sum_{j=1}^{n} a_j e^{ij\lambda} \right|^2 dF(\lambda) \geq 0.$$

$$\Rightarrow \gamma \text{ positive semidefinite.}$$

2) "$\Rightarrow$":

$$0 \leq f_n(\lambda) = \frac{1}{2\pi n} \sum_{j,l=1}^{n} e^{-ij\lambda} \gamma (j-l) e^{il\lambda} = \frac{1}{2\pi} \sum_{s=-(n-1)}^{n-1} \left(1 - \frac{|s|}{n}\right) e^{-is\lambda} \gamma (s)$$

Then

$$F_n(\lambda) = 0 \; (\lambda \leq -\pi), \; F_n(\lambda) = F_n(\pi) \;\; (\lambda > \pi),$$

$$F_n(\lambda) = \int_{-\pi}^{\lambda} f_n(\nu) \, d\nu \; (-\pi < \lambda \leq \pi)$$

is a sequence of gcdf with

$$\gamma_n(k) = \int_{-\pi}^{\pi} e^{ik\lambda} dF_n(\lambda) = \frac{1}{2\pi} \sum_{s=-(n-1)}^{n-1} \left(1 - \frac{|s|}{n}\right) \gamma (s) \int_{-\pi}^{\pi} e^{i(k-s)\lambda} d\lambda,$$

and hence

$$\gamma_n(k) = \left(1 - \frac{|k|}{n}\right) \gamma (k) \;\; (|k| \leq n-1), \; \gamma_n(k) = 0 \; (|k| \geq n).$$

Therefore,

$$\forall k \in \mathbb{Z} : \lim_{n \to \infty} \gamma_n(k) = \lim_{n \to \infty} \int_{-\pi}^{\pi} e^{ik\lambda} dF_n(\lambda) = \gamma (k).$$

Remaining question:

$$\exists \text{ gcdf } F \text{ s.t. } \gamma (k) = \int e^{ik\lambda} dF(\lambda) \, ?$$

Lemma 3.10 together with

$$0 \leq F_n(\lambda) \leq F_n(\pi) = \frac{1}{2\pi n} \sum_{s=-(n-1)}^{n-1} (n - |s|)\, \gamma(s) \int_{-\pi}^{\pi} e^{-is\lambda} d\lambda = \gamma(0) < \infty$$

implies

$$\exists \text{ gcdf } F \text{ and } F_{n_j}\ (j \in \mathbb{N}) \text{ s.t. } F_{n_j} \to F.$$

Then

$$\forall g \in C[-\pi, \pi] : \lim_{j\to\infty} \int g(\lambda)\, dF_{n_j}(\lambda) = \int g(\lambda)\, dF(\lambda),$$

and in particular

$$\lim_{j\to\infty} \int e^{ik\lambda} dF_{n_j}(\lambda) = \int e^{ik\lambda} dF(\lambda).$$

Hence,

$$\forall k \in \mathbb{Z} : \gamma(k) = \lim_{j\to\infty} \int e^{ik\lambda} dF_{n_j}(\lambda) = \int e^{ik\lambda} dF(\lambda).$$

Remark 3.4 Theorem 3.4 is known as Herglotz's theorem which is a special case of Bochner's theorem.

Corollary 3.2

$$\exists X_t \in \mathbb{C}\ (t \in \mathbb{Z}) \text{ weakly stationary s.t. } \forall k \in \mathbb{Z} : cov(X_t, X_{t+k}) = \gamma(k)$$

$$\Leftrightarrow$$

$$\gamma : \mathbb{Z} \to \mathbb{C} \text{ Hermitian positive semidefinite}$$

$$\Leftrightarrow$$

$$\exists F \text{ gcdf s.t. } F(-\pi) = 0,\ F(\pi) = F(\infty),\ \gamma(k) = \int_{-\pi}^{\pi} e^{ik\lambda} dF(\lambda)\ \ (k \in \mathbb{Z})$$

Corollary 3.3

$$\exists X_t \in \mathbb{R}\ (t \in \mathbb{Z}) \text{ weakly stationary s.t. } \forall k \in \mathbb{Z} : cov(X_t, X_{t+k}) = \gamma(k)$$

$$\Leftrightarrow$$

$$\gamma : \mathbb{Z} \to \mathbb{R} \text{ symmetric positive semidefinite}$$

$$\Leftrightarrow$$

$$\gamma(k) = \gamma(-k)$$

and

$$\exists F \text{ gcdf s.t.:}$$

$$F(-\pi) = 0,\ F(\pi) = F(\infty),$$

$$\gamma(k) = \int_{-\pi}^{\pi} e^{ik\lambda} dF(\lambda) \ (k \in \mathbb{Z}).$$

Definition 3.6 Let X_t $(t \in \mathbb{Z})$ be stationary with autocovariance function

$$\gamma_X(k) = \int_{-\pi}^{\pi} e^{ik\lambda} dF(\lambda) \ (k \in \mathbb{Z}),$$

and

$$F \text{ gcdf s.t. } F(-\pi) = 0.$$

Then F is called spectral distribution of X_t. If the derivative $f = F'$ exists, then f is called spectral density of X_t.

Corollary 3.4 *Assumptions:*

$$X_t \in \mathbb{R} \ (t \in \mathbb{Z}) \text{ weakly stationary}$$

$$\gamma_X(k) = \int_{-\pi}^{\pi} e^{ik\lambda} dF(\lambda)$$

$$\exists f \text{ s.t. } \forall \lambda \in [-\pi, \pi] : F(\lambda) = \int_{-\pi}^{\lambda} f(\nu)\, d\nu$$

Then,

$$\text{Lebesgue-a.e. in } [-\pi, \pi]:$$

$$\exists f(\lambda) = F'(\lambda) \text{ and } f(\lambda) \geq 0.$$

Proof

$$F \text{ nondecreasing} \Rightarrow f = F' \geq 0$$

3.4.3 Calculation and Properties of F and f

Problem 3.6 Calculation of F

Solution 3.6 Let

$$L^2[-\pi,\pi] = \left\{ g \mid g : [-\pi,\pi] \to \mathbb{C} \text{ s.t. } \|g\|^2_{L^2[-\pi,\pi]} = \int_{-\pi}^{\pi} |g(\lambda)|^2 \, d\lambda < \infty \right\},$$

$$\langle g, h \rangle_{L^2[-\pi,\pi]} = \int_{-\pi}^{\pi} g(\lambda) \overline{h(\lambda)} d\lambda \ (g, h \in L^2[-\pi,\pi]),$$

$$e_k(\lambda) = e^{ik\lambda} \ (k \in \mathbb{Z}),$$

$$h_{\lambda_1,\lambda_2}(\lambda) = 1\{\lambda_1 < \lambda \leq \lambda_2\}.$$

Then

$$F_X(\lambda_2) - F_X(\lambda_1) = \int_{-\pi}^{\pi} h_{\lambda_1,\lambda_2}(\lambda) \, dF_X(\lambda),$$

$$e_k \ (k \in \mathbb{Z}) \text{ orthogonal basis in } L^2[-\pi,\pi],$$

$$h_{\lambda_1,\lambda_2} \in L^2[-\pi,\pi].$$

Hence,

$$h_{\lambda_1,\lambda_2}(\lambda) \underset{L^2[-\pi,\pi]}{=} \sum_{k=-\infty}^{\infty} a_k e_k(\lambda)$$

where

$$a_k = a_k(\lambda_1,\lambda_2) = \frac{1}{2\pi} \int_{-\pi}^{\pi} e^{-ik\lambda} h_{\lambda_1,\lambda_2}(\lambda) \, d\lambda$$

We will show

$$F_X(\lambda_2) - F_X(\lambda_1) = \sum_{k=-\infty}^{\infty} a_k \gamma_X(k). \tag{3.1}$$

To obtain a general answer some auxiliary results are needed.

Lemma 3.11

$$F \; gcdf, \; g \in L^2(F) \Rightarrow g \in L^1(F)$$

Proof

$$\begin{aligned} \|g\|_{L^1(F)} &= \int |g(x)| \, dF(x) = \langle |g|, 1 \rangle_{L^2(F)} \\ &\leq \|g\|_{L^2(F)} \|1\|_{L^2(F)} = \|g\|_{L^2(F)} F(\infty) < \infty \end{aligned}$$

Lemma 3.12 *Assumptions:*

$$F \; gcdf, \; g, g_n \in L^2(F)$$

$$\lim_{n \to \infty} \|g_n - g\|_{L^2(F)} = 0$$

Then

$$\lim_{n \to \infty} \|g_n - g\|_{L^1(F)} = 0$$

Proof

$$\begin{aligned} \int |g_n(x) - g(x)| \, dF(x) &= \langle |g_n - g|, 1 \rangle_{L^2(F)} \\ &\leq \|g_n - g\|_{L^2(F)} F(\infty) \to 0 \end{aligned}$$

Corollary 3.5 *Assumptions:*

$$F \; gcdf, \; g, g_n \in L^2(F)$$

$$\lim_{n \to \infty} \|g_n - g\|_{L^2(F)} = 0$$

Then

$$\lim_{n \to \infty} \int g_n(x) \, dF(x) = \int g(x) \, dF(x)$$

Proof

$$\begin{aligned} \left| \int g_n(x) \, dF(x) - \int g(x) \, dF(x) \right| &\leq \int |g_n(x) - g(x)| \, dF(x) \\ &= \|g_n - g\|_{L^1(F)} \to 0 \end{aligned}$$

Problem 3.7 To derive (3.1), the following question has to be answered first: Does

$$h_{\lambda_1,\lambda_2} \underset{L^2[-\pi,\pi]}{=} \sum_{k=-\infty}^{\infty} a_k e_k$$

imply

$$h_{\lambda_1,\lambda_2} \underset{L^2(F_X)}{=} \sum_{t=-\infty}^{\infty} a_t e_t?,$$

i.e.

$$h_n = \sum_{k=-n}^{n} a_k e_k \underset{L^2(F_X)}{\rightarrow} h_{\lambda_1,\lambda_2}?$$

Solution 3.7 Application of the Dirichlet kernel.

Definition 3.7 The sequence of functions

$$D_n : \mathbb{R} \to \mathbb{R},\ \lambda \to D_n(\lambda) = \sum_{k=-n}^{n} e^{ik\lambda} \ (n \in \mathbb{N})$$

is called Dirichlet kernel.

Lemma 3.13

$$D_n(2\pi j) = 2n + 1,$$

$$D_n(\lambda) = \sum_{k=-n}^{n} e^{ik\lambda} = \frac{\sin\left(n + \frac{1}{2}\right)\lambda}{\sin\frac{\lambda}{2}} \ (\lambda \neq 2\pi j).$$

Proof

1)

$$D_n(2\pi j) = \sum_{k=-n}^{n} 1 = 2n + 1$$

2) For $\lambda \neq 2\pi j$,

$$D_n(\lambda) = \sum_{k=-n}^{n} e^{ik\lambda} = 1 + \sum_{k=1}^{n} \left(e^{i\lambda}\right)^k + \sum_{k=1}^{n} \left(e^{-i\lambda}\right)^k$$

$$= \frac{\exp(i\lambda(n+1)) - \exp(-i\lambda n)}{\exp(i\lambda) - 1}$$

$$= \frac{\exp\left(i\lambda\left(n+\frac{1}{2}\right)\right) - \exp\left(-i\lambda\left(n+\frac{1}{2}\right)\right)}{\exp\left(i\frac{1}{2}\lambda\right) - \exp\left(-i\frac{1}{2}\lambda\right)}$$

$$= \frac{\sin\left(n+\frac{1}{2}\right)\lambda}{\sin\frac{\lambda}{2}}.$$

Lemma 3.14 *Assumptions:*

$$J_n(x) = \int_0^x D_n(\lambda)\, d\lambda$$

$$0 < \delta < 2\pi$$

Then

$$\lim_{n\to\infty} \sup_{x\in[\delta,2\pi-\delta]} |J_n(x) - \pi| = 0$$

and

$$\exists M \in \mathbb{R} \text{ s.t. } \forall n \in \mathbb{N} : \sup_{x\in[0,2\pi-\delta]} |J_n(x)| \le M.$$

Proof See Brockwell and Davis (1991, p. 157).

Theorem 3.5 *Assumptions:*

$$-\pi < \lambda_1 < \lambda_2 < \pi$$

$$h_{\lambda_1,\lambda_2}(\lambda) = 1\{\lambda_1 < \lambda \le \lambda_2\}$$

$$a_k = \frac{1}{2\pi}\int_{-\pi}^{\pi} e^{-ik\lambda} h_{\lambda_1,\lambda_2}(\lambda)\, d\lambda$$

$$h_n = \sum_{k=-n}^{n} a_k e_k$$

$$U \subseteq [-\pi,\pi],\ U \text{ open},\ \lambda_1,\lambda_2 \in U$$

Then

$$\lim_{n\to\infty} \sup_{\lambda\in[-\pi,\pi]\setminus U} |h_n(\lambda) - h_{\lambda_1,\lambda_2}(\lambda)| = 0$$

and

$$\exists M \in \mathbb{R} \text{ s.t. } \sup_{\lambda \in [-\pi, \pi]} |h_n(\lambda)| \leq M.$$

Proof See Brockwell and Davis (1991, p. 158).

Lemma 3.15 *Assumptions:*

$$h : \mathbb{R} \to \mathbb{R}$$

$$h(\lambda + 2\pi j) = h(\lambda) \ (\lambda \in [-\pi, \pi], j \in \mathbb{Z})$$

$$h : [-\pi, \pi] \to \mathbb{R} \text{ of bounded variation}$$

Then

$$\forall \lambda \in \mathbb{R} : \lim_{n \to \infty} \frac{1}{2\pi} \int_{-\pi}^{\pi} D_n(u) h(\lambda - u) \, du = \frac{h(\lambda^+) + h(\lambda^-)}{2}$$

Proof See e.g. Edwards (1967, p. 150).

Corollary 3.6 *Assumptions:*

$$X_t \in \mathbb{R} \ (t \in \mathbb{Z}) \text{ weakly stationary}$$

$$\gamma_X(k) = \int_{-\pi}^{\pi} e^{ik\lambda} dF_X(\lambda)$$

$$-\pi < \lambda_1 < \lambda_2 < \pi$$

$$\lambda_1, \lambda_2 \in J_{cont}(F_X)$$

$$h_{\lambda_1, \lambda_2}, h_n \text{ as in Theorem 3.5.}$$

Then

$$h_n \underset{L^2(F_X)}{\to} h_{\lambda_1, \lambda_2}$$

Proof

1) Let

$$U_\delta = (\lambda_1 - \delta, \lambda_1 + \delta) \cup (\lambda_2 - \delta, \lambda_2 + \delta) \subset [-\pi, \pi].$$

Then

$$\|h_n - h_{\lambda_1,\lambda_2}\|^2_{L^2(F_X)} = \Delta_{1,n}(\delta) + \Delta_{2,n}(\delta)$$

with

$$\Delta_{1,n}(\delta) = \int_{[-\pi,\pi]\setminus U_\delta} |h_n(\lambda) - h_{\lambda_1,\lambda_2}(\lambda)|^2 \, dF_X(\lambda),$$

$$\Delta_{2,n}(\delta) = \int_{U_\delta} |h_n(\lambda) - h_{\lambda_1,\lambda_2}(\lambda)|^2 \, dF_X(\lambda).$$

2)

$$\Delta_{1,n}(\delta) \le 2\pi \sup_{\lambda\in[-\pi,\pi]\setminus U_\delta} |h_n(\lambda) - h_{\lambda_1,\lambda_2}(\lambda)|^2,$$

and Theorem 3.5 imply

$$\lim_{n\to\infty} \Delta_{1,n}(\delta) = 0.$$

3)

$$\Delta_{2,n}(\delta) \le \sum_{j=1}^{2} \int_{\lambda_j-\delta}^{\lambda_j+\delta} |h_n(\lambda) - h_{\lambda_1,\lambda_2}(\lambda)|^2 \, dF_X(\lambda),$$

$$\int_{\lambda_j-\delta}^{\lambda_j+\delta} |h_n(\lambda) - h_{\lambda_1,\lambda_2}(\lambda)|^2 \, dF_X(\lambda) = I_{j,1,n}(\delta) + I_{j,2,n}(\delta)$$

where

$$I_{j,1,n}(\delta) = \int_{\lambda\in[\lambda_j-\delta,\lambda_j)\cup(\lambda_j,\lambda_j+\delta]} |h_n(\lambda) - h_{\lambda_1,\lambda_2}(\lambda)|^2 \, dF_X(\lambda),$$

$$I_{j,2,n} = \left|h_n(\lambda_j) - h_{\lambda_1,\lambda_2}(\lambda_j)\right|^2 \left[F_X(\lambda_j) - F_X(\lambda_j^-)\right].$$

Theorem 3.5 implies

$$\sup_{\lambda\in[-\pi,\pi]} |h_n(\lambda)| \le M$$

and hence

$$\sup_{\lambda\in[-\pi,\pi]} |h_n(\lambda) - h_{\lambda_1,\lambda_2}(\lambda)|^2 \le (M+1)^2$$

and

$$I_{j,1,n}(\delta) \le (M+1)^2 \left\{ F_X\left(\lambda_j^-\right) - F_X\left(\lambda_j - \delta\right) + F_X\left(\lambda_j + \delta\right) - F_X\left(\lambda_j\right) \right\}.$$

Then

$$\lim_{\delta \downarrow 0} F_X\left(\lambda_j - \delta\right) = F_X\left(\lambda_j^-\right), \ \lim_{\delta \downarrow 0} F_X\left(\lambda_j + \delta\right) = F_X\left(\lambda_j\right)$$

lead to

$$\lim_{\delta \downarrow 0} I_{j,1,n}(\delta) = 0.$$

For $I_{j,2,n}$,

$$\left|h_n(\lambda) - h_{\lambda_1,\lambda_2}(\lambda)\right|^2 \le (M+1)^2$$

and

$$\lambda_1, \lambda_2 \in J_{cont}(F_X) \Rightarrow F_X\left(\lambda_j\right) - F_X\left(\lambda_j^-\right) = 0$$

so that

$$I_{j,2,n} = 0.$$

4) From (1) to (3) we obtain

$$\lim_{n \to \infty} \left\| h_n - h_{\lambda_1,\lambda_2} \right\|_{L^2(F_X)} = 0.$$

We now can prove the following inversion formula:

Corollary 3.7 *Assumptions:*

$$X_t \in \mathbb{R} \ (t \in \mathbb{Z}) \text{ weakly stationary}$$

$$\gamma_X(k) = \int_{-\pi}^{\pi} e^{ik\lambda} dF_X(\lambda)$$

$$-\pi < \lambda_1 < \lambda_2 < \pi$$

$$\lambda_1, \lambda_2 \in J_{cont}(F_X)$$

$$h_{\lambda_1,\lambda_2} \text{ as in Theorem 3.5}$$

$$a_k = \frac{1}{2\pi} \int_{-\pi}^{\pi} e^{-ik\lambda} h_{\lambda_1,\lambda_2}(\lambda)\, d\lambda.$$

Then

$$F_X(\lambda_2) - F_X(\lambda_1) = \sum_{k=-\infty}^{\infty} a_k \gamma_X(k).$$

Proof

$$h_{\lambda_1,\lambda_2}(\lambda) = 1\{\lambda_1 < \lambda \le \lambda_2\}, h_n = \sum_{k=-n}^{n} a_k e^{-ik\lambda} = \sum_{k=-n}^{n} a_k e_k,$$

$$a_k = \frac{1}{2\pi} \int_{-\pi}^{\pi} e^{-ik\lambda} h_{\lambda_1,\lambda_2}(\lambda)\, d\lambda,$$

$$h_n \underset{L^2(F_X)}{\rightarrow} h_{\lambda_1,\lambda_2}.$$

Hence

$$\lim_{n\to\infty} \int_{-\pi}^{\pi} h_n(\lambda)\, dF_X(\lambda) = \int_{-\pi}^{\pi} h_{\lambda_1,\lambda_2}(\lambda)\, dF_X(\lambda) = F_X(\lambda_2) - F_X(\lambda_1).$$

Now

$$\lim_{n\to\infty} \int_{-\pi}^{\pi} h_n(\lambda)\, dF_X(\lambda) = \lim_{n\to\infty} \sum_{k=-n}^{n} a_k \int_{-\pi}^{\pi} e^{ik\lambda} dF_X(\lambda)$$
$$= \sum_{k=-\infty}^{\infty} a_k \gamma_X(k)$$

so that

$$F_X(\lambda_2) - F_X(\lambda_1) = \sum_{k=-\infty}^{\infty} a_k \gamma_X(k).$$

F_X is symmetric in the following sense:

Corollary 3.8 *Assumptions:*

$$X_t \in \mathbb{R}\ (t \in \mathbb{Z}) \text{ weakly stationary}$$

$$\gamma_X(k) = \int_{-\pi}^{\pi} e^{ik\lambda} dF_X(\lambda)$$

Then

$$\forall \lambda \in (0,\pi) : F_X(\lambda) = F_X(\pi^-) - F_X(-\lambda^-)$$

Proof Let

$$\lambda_0 \in [0, \pi),$$

and

$$A_\delta = \{-\pi + \delta, -\lambda_0 - \delta, \lambda_0 + \delta, \pi - \delta\} \ (\delta > 0).$$

Then

$$C_{jump}(F) \text{ countable}$$

implies

$$\exists \delta_N > 0 \ (N \in \mathbb{N}) \text{ s.t.:}$$

$$\lim_{N\to\infty} \delta_N = 0, \lambda_0 + \delta_N < \pi - \delta_N \text{ and } A_{\delta_N} \subseteq C_{cont}(F).$$

Define

$$h_N(\lambda) = 1\{-\pi + \delta_N < \lambda \leq \lambda_0 + \delta_N\} - 1\{-\lambda_0 - \delta_N < \lambda \leq \pi - \delta_N\}.$$

Now

$$h_N \text{ continuous for } \lambda \notin A_{\delta_N}, A_{\delta_N} \subseteq C_{cont}(F)$$

implies

$$\int_{-\pi}^{\pi} h_N(\lambda)\, dF_X(\lambda) = \int_{[-\pi,\pi]\setminus A_{\delta_N}} h_N(\lambda)\, dF_X(\lambda).$$

Then

$$\forall \lambda \in [-\pi, \pi]\setminus A_{\delta_N} : h_N(\lambda) = -h_N(-\lambda) \text{ and } h_N \text{ continuous at } \lambda$$

implies

$$\exists b_k \in \mathbb{R} \ (k \geq 1) \text{ s.t. for } h_{n,N}(\lambda) = \sum_{k=1}^{n} b_k \sin k\lambda :$$

$$h_N(\lambda) = \lim_{n\to\infty} h_{n,N}(\lambda) \ \ (\forall \lambda \in [-\pi, \pi]\setminus A_{\delta_N}).$$

From Corollary 3.6, we also have

$$h_{n,N} \underset{L^2(F_X)}{\rightarrow} h_N \ (n \rightarrow \infty).$$

Now

$$\gamma_X(k) = \int_{-\pi}^{\pi} \cos(k\lambda)\, dF_X(\lambda) + i \int_{-\pi}^{\pi} \sin(k\lambda)\, dF_X(\lambda) \in \mathbb{R}$$

so that

$$\int_{-\pi}^{\pi} \sin(k\lambda)\, dF_X(\lambda) = 0 \ (k \in \mathbb{Z}).$$

Therefore

$$\int_{[-\pi,\pi]\setminus A_{\delta_N}} h_N(\lambda)\, dF_X(\lambda) = \lim_{n\to\infty} \int_{[-\pi,\pi]\setminus A_{\delta_N}} h_{n,N}(\lambda)\, dF_X(\lambda) = 0.$$

Hence, for all $N \in \mathbb{N}$,

$$\begin{aligned} 0 &= \int_{-\pi}^{\pi} h_N(\lambda)\, dF_X(\lambda) \\ &= F_X(\lambda_0 + \delta_N) - F_X(-\pi + \delta_N) - [F_X(\pi - \delta_N) - F_X(-\lambda_0 - \delta_N)], \end{aligned}$$

and using

$$\lim_{\delta_N \downarrow 0} F_X(\lambda + \delta_N) = F_X(\lambda), \ F_X(-\pi) = 0$$

we obtain

$$F_X(\lambda_0) = F_X(\pi^-) - F_X\left(-\lambda_0^-\right).$$

Corollary 3.9 *Assumptions:*

$$X_t \in \mathbb{R} \ (t \in \mathbb{Z}) \text{ weakly stationary}$$

$$\gamma_X(k) = \int_{-\pi}^{\pi} e^{ik\lambda} dF_X(\lambda)$$

$$F_X(\lambda) = \int_{-\pi}^{\lambda} f_X(\lambda)\, d\lambda \ (\lambda \in [-\pi, \pi]).$$

Then

$$f(\lambda) = f(-\lambda) \text{ (Lebesgue-a.e. in } [-\pi, \pi]\text{)}$$

Proof

$$F(\lambda) = \int_{-\pi}^{\lambda} f(\lambda)\, d\lambda$$

implies

$$F(\lambda) = F(\pi) - F(-\lambda)$$

and

$$f(\lambda) = \frac{d}{d\lambda}\left[F(\pi) - F(-\lambda)\right] = f(-\lambda).$$

Lemma 3.16 *Assumptions:*

$$F, G = \text{spectral distribution functions}$$

$$\gamma_F(k) = \int_{-\pi}^{\pi} e^{ik\lambda} dF(\lambda),\ \gamma_G(k) = \int_{-\pi}^{\pi} e^{ik\lambda} dG(\lambda)\ (k \in \mathbb{Z}).$$

Then

$$\forall k \in \mathbb{Z} : \gamma_F(k) = \gamma_G(k) \Leftrightarrow \forall \lambda \in [-\pi, \pi] : F(\lambda) = G(\lambda)$$

Proof Analogous arguments as in Corollaries 3.7 and 3.8.

Problem 3.8 In general, proving the existence of F given a function γ is not always easy. Are there simpler sufficient conditions for positive semidefiniteness of γ?

Solution 3.8 Spectral density $f \geq 0$.

Definition 3.8 Given a function

$$\gamma : \mathbb{Z} \to \mathbb{R},$$

its discrete Fourier transform $\mathcal{F}_\gamma$ is defined by

$$\mathcal{F}_\gamma : [-\pi, \pi] \to \mathbb{R}, \lambda \to \mathcal{F}_\gamma(\lambda) = \frac{1}{2\pi} \sum_{k=-\infty}^{\infty} e^{-ik\lambda} \gamma(k).$$

Corollary 3.10 *Assumptions:*

$$\gamma : \mathbb{Z} \to \mathbb{R} \text{ symmetric}$$

$$\sum_{k=-\infty}^{\infty} |\gamma(k)| < \infty$$

Then

$$\exists X_t \in \mathbb{R} \ (t \in \mathbb{Z}) \text{ weakly stationary s.t. } cov\,(X_t, X_{t+k}) = \gamma(k) \ \ (k \in \mathbb{Z})$$

$$\Leftrightarrow$$

$$\forall \lambda \in [-\pi, \pi] : \mathscr{F}_{\gamma}(\lambda) \geq 0.$$

Moreover, if this holds, then

$$\gamma(k) = \int_{-\pi}^{\pi} e^{ik\lambda} f(\lambda)\, d\lambda \text{ with } f(\lambda) = \mathscr{F}_{\gamma}(\lambda).$$

Proof

1) "⇒" Let

$$cov\,(X_t, X_{t+k}) = \gamma(k) \ \ (k \in \mathbb{Z}),$$

and define

$$\begin{aligned} f_n(\lambda) &= \frac{1}{2\pi n} \sum_{j,l=1}^{n} e^{-ij\lambda} \gamma(j-l)\, e^{il\lambda} \\ &= \frac{1}{2\pi} \sum_{k=-(n-1)}^{n-1} \left(1 - \frac{|k|}{n}\right) e^{-ik\lambda} \gamma(k). \end{aligned}$$

a)

$$\sum_{k=-\infty}^{\infty} |\gamma(k)| < \infty \Rightarrow \forall \lambda \in [-\pi, \pi] : \exists \mathscr{F}_{\gamma}(\lambda) \in \mathbb{R}$$

b)

$$\gamma \text{ positive semidefinite} \Rightarrow f_n(\lambda) \geq 0$$

c)

$$\sum_{k=-\infty}^{\infty} |\gamma(k)| < \infty$$

$$\Rightarrow \lim_{k\to\infty} |k\gamma(k)| = 0$$

$$\Rightarrow |k\gamma(k)| \;\; (k \in \mathbb{N}) \text{ Cesaro summable with limit } 0$$

Hence,

$$\lim_{n\to\infty} \left| f_n(\lambda) - \frac{1}{2\pi} \sum_{k=-(n-1)}^{n-1} e^{-ik\lambda}\gamma(k) \right| = \frac{1}{2\pi} \lim_{n\to\infty} \left| \sum_{k=-(n-1)}^{n-1} \frac{k}{n}\gamma(k) \right|$$
$$\leq \frac{1}{2\pi} \lim_{n\to\infty} \sum_{k=-(n-1)}^{n-1} \left|\frac{k}{n}\right| |\gamma(k)| = 0$$

and

$$0 \leq \lim_{n\to\infty} f_n(\lambda) = \frac{1}{2\pi} \sum_{k=-\infty}^{\infty} e^{-ik\lambda}\gamma(k) = \mathscr{F}_\gamma(\lambda).$$

From the proof of Theorem 3.4 we have

$$\gamma(k) = \int_{-\pi}^{\pi} e^{ik\lambda} dF(\lambda) \text{ with } F(\lambda) = \int_{-\pi}^{\lambda} \mathscr{F}_\gamma(\nu)\, d\nu$$

and hence

$$f(\lambda) = F'(\lambda) = \mathscr{F}_\gamma(\lambda).$$

2) "$\Leftarrow$"

$$\sum_{k=-\infty}^{\infty} |\gamma(k)| < \infty, \mathscr{F}_\gamma(\lambda) \geq 0$$

a) Define

$$F(\lambda) = 0\,(\lambda < -\pi), F(\lambda) = \int_{-\pi}^{\lambda} \mathscr{F}_\gamma(\nu)\, d\nu = F(\infty) \;\; (\lambda \in [-\pi, \pi]).$$

Then

$$0 \leq \mathcal{F}_{\gamma}(\lambda) \leq \frac{1}{2\pi} \sum_{k=-\infty}^{\infty} |\gamma(k)| < \infty \Rightarrow F = \text{spectral distribution}$$

b)

$$\left| \frac{1}{2\pi} \sum_{k=-n}^{n} e^{-ik\lambda} \gamma(k) \right| \leq \frac{1}{2\pi} \sum_{k=-\infty}^{\infty} |\gamma(k)| < \infty$$

implies

$$\begin{aligned} \int_{-\pi}^{\pi} e^{ik\lambda} dF(\lambda) &= \int_{-\pi}^{\pi} e^{ik\lambda} \mathcal{F}_{\gamma}(\lambda)\, d\lambda \\ &= \int_{-\pi}^{\pi} e^{ik\lambda} \lim_{n\to\infty} \left(\frac{1}{2\pi} \sum_{m=-n}^{m} e^{-im\lambda} \gamma(m) \right) d\lambda \\ &= \lim_{n\to\infty} \sum_{m=-n}^{n} \gamma(m) \left(\frac{1}{2\pi} \int_{-\pi}^{\pi} e^{i(k-m)\lambda} d\lambda \right) \\ &= \frac{1}{2\pi} \sum_{u=-\infty}^{\infty} \gamma(m)\, 2\pi \delta_{km} = \gamma(k)\,. \end{aligned}$$

Hence, γ is an acf, and

$$\gamma(k) = \int_{-\pi}^{\pi} e^{ik\lambda} f(\lambda)\, d\lambda \text{ with } f(\lambda) = \mathcal{F}_{\gamma}(\lambda)\,.$$

Remark 3.5 The same theorem holds for $\gamma : \mathbb{Z} \to \mathbb{C}$ with $\gamma(-k) = \overline{\gamma(k)}$.

Example 3.9

$$\gamma(k) = \delta_{0k}\sigma^2 < \infty \;(k \in \mathbb{Z})$$

$$\Rightarrow f(\lambda) = \frac{\sigma^2}{2\pi} \geq 0 \Rightarrow \gamma \text{ is an acf}$$

Example 3.10

$$\gamma(0) = \sigma^2 \geq 0,\ \gamma(\pm 1) = \sigma^2 \rho,\ \gamma(k) = 0\ (|k| \geq 2)$$

$$\Rightarrow f(\lambda) = \frac{\sigma^2}{2\pi}(1 + 2\rho \cos k\lambda)$$

Therefore

$$\forall\lambda \in [-\pi, \pi] : f(\lambda) \geq 0 \Leftrightarrow |\rho| \leq \frac{1}{2}.$$

Example 3.11

$$0 < \sigma_\varepsilon^2 < \infty,\ |\varphi| < 1, \gamma(k) = \frac{\varphi^{|k|}}{1 - \varphi^2}$$

$$\Rightarrow f(\lambda) = \frac{\sigma_\varepsilon^2}{2\pi} \sum_{k=-\infty}^{\infty} \varphi^{|k|} e^{ik\lambda} = \frac{\sigma_\varepsilon^2}{2\pi} \frac{1}{\left|1 - \varphi e^{i\lambda}\right|^2} > 0.$$

Chapter 4
Spectral Representation of Univariate Time Series

4.1 Motivation

Problem 4.1

$$X_t \text{ second order stationary}, E(X_t) = 0, \gamma_X(k) = \int_{-\pi}^{\pi} e^{ik\lambda} dF_X(\lambda)$$

$$\Rightarrow \text{additive decomposition of } X_t \text{ into periodic components?}$$

Solution 4.1

$$X_t = \int_{-\pi}^{\pi} e^{it\lambda} dZ_X(\lambda) \tag{4.1}$$

$$cov(dZ_X(\lambda_1), dZ_X(\lambda_2)) = 0 \ (\lambda_1 \neq \lambda_2)$$

$$var(dZ_X(\lambda)) = E\left[|dZ_X(\lambda)|^2\right] = dF_X(\lambda)$$

Interpretation:

$$dZ_X(\lambda) = \text{random amplitude at frequency } \lambda$$

$$\gamma_X(0) = \int_{-\pi}^{\pi} dF_X(\lambda) = \int_{-\pi}^{\pi} E\left[|dZ_X(\lambda)|^2\right]$$

J. Beran, *Mathematical Foundations of Time Series Analysis*,
https://doi.org/10.1007/978-3-319-74380-6_4

4.2 Harmonic Processes

Representation (4.1) is first derived for a class of simple processes.

Definition 4.1 Let

$$A_j \in \mathbb{C}, E\left(A_j\right) = 0, cov\left(A_j, A_l\right) = E\left(A_j \overline{A_l}\right) = \sigma_j^2 \delta_{jl} < \infty,$$

$$-\pi < \lambda_1 < \lambda_2 < \ldots < \lambda_k \leq \pi.$$

Then

$$X_t = \sum_{j=1}^{k} A e^{it\lambda_j} \ (t \in \mathbb{Z})$$

is called a harmonic process.

Definition 4.2 Let

$$Z(\lambda) \in \mathbb{C}, E[Z(\lambda)] = 0, E\left[|Z(\lambda)|^2\right] < \infty \ (\lambda \in [-\pi, \pi]).$$

Then Z is called a right-continuous stochastic process, if

$$\forall \lambda \in [-\pi, \pi) : \lim_{\delta \downarrow 0} \|Z(\lambda + \delta) - Z(\lambda)\|_{L^2(\Omega;\mathbb{C})}^2 = \lim_{\delta \downarrow 0} E\left[|Z(\lambda + \delta) - Z(\lambda)|^2\right] = 0.$$

Moreover, Z is called an orthogonal increment process, if

$$\forall \lambda_1 < \lambda_2 \leq \lambda_3 < \lambda_4 : \langle Z(\lambda_2) - Z(\lambda_1), Z(\lambda_4) - Z(\lambda_3) \rangle = 0$$

Lemma 4.1 *Assumption:*

$$X_t = \sum_{j=1}^{k} A_j e^{it\lambda_j} = \textit{harmonic process}.$$

Then

$$X_t = \textit{weakly and strictly stationary}$$

Proof Follows directly from the definition.

Lemma 4.2 *Assumptions:*

$$X_t = \sum_{j=1}^{k} A_j e^{it\lambda_j} = \textit{harmonic process},$$

$$Z_X(\lambda) = \sum_{\lambda_j \le \lambda} A_j,\ F_X(\lambda) = \sum_{\lambda_j \le \lambda} \sigma_j^2.$$

Then

$$Z_X(\lambda) = \textit{right-continuous orthogonal increment process},$$

$$X_t = \int_{-\pi}^{\pi} e^{it\lambda_j} dZ(\lambda) \ \ (t \in \mathbb{Z})$$

and

$$\gamma_X(k) = cov(X_t, X_{t+k}) = \int_{-\pi}^{\pi} e^{ik\lambda} dF_X(\lambda) \ \ (k \in \mathbb{Z}).$$

Proof

1)

$$Z_X(\lambda) = \sum_{\lambda_j \le \lambda} A_j \Rightarrow Z_X(\lambda) \text{ right-continuous}$$

2)

$$E[Z_X(\lambda)] = \sum_{\lambda_j \le \lambda} E[A_j] = 0.$$

3) Let

$$-\pi \le \nu_1 < \nu_2 \le \nu_3 < \nu_4 < \pi.$$

Then

$$\langle Z(\nu_2) - Z(\nu_1), Z(\nu_4) - Z(\nu_3)\rangle = \sum_{j:\nu_1 < \lambda_j \le \nu_2} \sum_{l:\nu_3 < \lambda \le \nu_4} E\left[A_j \overline{A_l}\right] = 0$$

4) Riemann-Stieltjes integral (pathwise)

$$\int_{-\pi}^{\pi} e^{it\lambda} dZ_X(\lambda) = \sum_{j=1}^{k} A_j e^{it\lambda_j}.$$

5)

$$cov(X_{t+k}, X_t) = E\left[\sum_{j=1}^{k} A_j e^{i(t+k)\lambda_j} \overline{\sum_{l=1}^{k} A_l e^{it\lambda_l}}\right]$$

$$= \sum_{j=1}^{k} e^{ik\lambda_j} \sigma_j^2 = \int_{-\pi}^{\pi} e^{ik\lambda} dF_X(\lambda).$$

Remark 4.1 Let X_t be a harmonic process. Then w.l.o.g.

$$X_t = \sum_{j=-k}^{k} A_j e^{it\lambda_j} \text{ with } \lambda_{-j} = -\lambda_j,$$

possibly with some $A_j \equiv 0$.

Lemma 4.3 *Assumptions:*

$$X_t = \sum_{j=-k}^{k} A_j e^{it\lambda_j} \ (t \in \mathbb{Z}) \text{ with } \lambda_{-j} = -\lambda_j,$$

$$X_t \in \mathbb{R}\ (t \in \mathbb{Z})$$

Then

$$A_j = a_j + ib_j,\ A_{-j} = a_j - ib_j\ (a_j, b_j \in \mathbb{R})$$

$$var(a_j) = var(b_j),\ cov(a_j, b_j) = 0.$$

Proof

$$A_j e^{it\lambda_j} = (a_j + ib_j)\, e^{it\lambda_j}$$
$$= (a_j \cos t\lambda_j - b_j \sin t\lambda_j) + i\,(a_j \sin t\lambda_j + b_j \cos t\lambda_j)$$

$$A_{-j} e^{-it\lambda_j} = (a_{-j} \cos t\lambda_j + b_{-j} \sin t\lambda_j) + i\,(-a_{-j} \sin t\lambda_j + b_{-j} \cos t\lambda_j).$$

1)

$$X_t \in \mathbb{R}\ (t \in \mathbb{Z})$$

$$\Rightarrow \operatorname{Im} A_j e^{it\lambda_j} = -\operatorname{Im} A_{-j} e^{-it\lambda_j}\ (\forall t \in \mathbb{Z})$$

$$\Rightarrow a_{-j} = a_j,\ b_{-j} = -b_j.$$

2) From 1),

$$\begin{aligned} 0 &= cov\left(A_j, A_{-j}\right) = E\left(A_j \overline{A_{-j}}\right) \\ &= E\left[\left(a_j + ib_j\right)\left(a_j + ib_j\right)\right] \\ &= var\left(a_j\right) - var\left(b_j\right) + i2cov\left(a_j, b_j\right), \end{aligned}$$

so that

$$var\left(a_j\right) = var\left(b_j\right),\ cov\left(a_j, b_j\right) = 0.$$

4.3 Extension to General Processes

Aim: "orthogonal" decomposition

$$X_t \underset{L^2(\Omega), a.s.}{=} \int_{-\pi}^{\pi} e^{it\lambda} dZ_X(\lambda).$$

Two questions: (a) given Z_X, define $\int g dZ_X$; (b) given X_t, find Z_X.

4.3.1 Stochastic Integrals with Respect to Z

Lemma 4.4 *Assumption:*

$$Z(\lambda)\ \ (\lambda \in [-\pi, \pi])\ \text{right-continuous orthogonal increment process.}$$

Then

$$\exists! gcdf\ F\ s.t.:$$

$$F(\lambda) = 0\ (\lambda \leq -\pi),\ F(\lambda) = F(\pi)\ \ (\lambda \geq \pi) \tag{4.2}$$

and

$$F(\lambda_2) - F(\lambda_1) = \|Z(\lambda_2) - Z(\lambda_1)\|^2_{L^2(\Omega)} \; (-\pi \le \lambda_1 \le \lambda_2 \le \pi) \tag{4.3}$$

Proof

1) Define

$$F(\lambda) = \|Z(\lambda) - Z(-\pi)\|^2_{L^2(\Omega)} \; (\lambda \in [-\pi, \pi]),$$

$$F(\lambda) = 0 \, (\lambda \le -\pi), \, F(\lambda) = F(\pi) \; (\lambda \ge \pi).$$

Then

$$F(\lambda_2) - F(\lambda_1) = \|Z(\lambda_2) - Z(\lambda_1)\|^2_{L^2(\Omega)} \; (-\pi \le \lambda_1 \le \lambda_2 \le \pi)$$

Also note that (4.2), (4.3) imply

$$F(\lambda) = \|Z(\lambda) - Z(-\pi)\|^2_{L^2(\Omega)} \; (\lambda \in [-\pi, \pi]).$$

2)

Z right-continuous process $\Rightarrow$ F right-continuous

3)

Z orthogonal increment process, $-\pi \le \lambda_1 \le \lambda_2 \le \pi$

$$\Downarrow$$

$$\begin{aligned} F(\lambda_2) &= \|Z(\lambda_2) - Z(\lambda_1) + Z(\lambda_1) - Z(-\pi)\|^2 \\ &= \|Z(\lambda_2) - Z(\lambda_1)\|^2 + \|Z(\lambda_1) - Z(-\pi)\|^2 \\ &\ge F(\lambda_1) \end{aligned}$$

4)

1),2),3) $\Rightarrow$ F gcdf with desired properties

Definition 4.3 Let

$$Z(\lambda)\ (\lambda \in [-\pi, \pi])\ \text{right-continuous orthogonal increment process},$$

$$F(\lambda) = 0\ (\lambda \le -\pi),\ F(\lambda) = F(\pi)\ \ (\lambda \ge \pi),$$

$$F(\lambda_2) - F(\lambda_1) = \|Z(\lambda_2) - Z(\lambda_1)\|^2_{L^2(\Omega)}\ \ (-\pi \le \lambda_1 \le \lambda_2 \le \pi).$$

Then F is called spectral distribution function associated with Z.

Problem 4.2 Let Z, F be as in Definition 4.3.

$$g \in L^2(F;\mathbb{C}) \Rightarrow \int g(\lambda)\, dZ(\lambda) = ?$$

Solution 4.2 Start with step functions, extension to $L^2(F;\mathbb{C})$ using an isomorphism.

Notation 10

$$\underline{\lambda} = (\lambda_1, \ldots, \lambda_m),\ -\pi \le \lambda_1 < \ldots < \lambda_{m+1} \le \pi$$

$$\mathbf{a} = (a_1, \ldots, a_m) \in \mathbb{C}^m$$

$$g_{\mathbf{a},\underline{\lambda}}(\lambda) = \sum_{k=1}^{m} a_k 1\{\lambda_k < \lambda \le \lambda_{k+1}\}$$

$$S = \left\{g_{\mathbf{a},\underline{\lambda}} \mid -\pi \le \lambda_1 < \ldots < \lambda_{m+1} \le \pi, \mathbf{a} \in \mathbb{C}^m\right\}$$
$$= \textit{set of step functions}$$

$$\langle g, h\rangle = \langle g, h\rangle_{L^2(F)} = \int g(\lambda)\, \overline{h(\lambda)} dF(\lambda),\ \|g\|^2 = \langle g, g\rangle$$

$$\overline{S}_F = L^2(F)\textit{-closure of } S$$

$$\xi_g = \sum_{k=1}^{m} a_k [Z(\lambda_{k+1}) - Z(\lambda_k)]\ \ (g = g_{\mathbf{a},\underline{\lambda}} \in S)$$

$$S_Z = S_Z(\Omega) = \left\{\xi_g : g \in S\right\}$$

$$\langle \xi, \eta\rangle = \langle \xi, \eta\rangle_{L^2(\Omega)} = cov(\xi, \eta),\ \|\xi\|^2 = \langle \xi, \xi\rangle = var(\xi)$$

$$\overline{S}_Z = L^2(\Omega)\textit{-closure of } S_Z$$

Definition 4.4 For

$$Z(\lambda)\ \ (\lambda\in[-\pi,\pi])\text{ right-continuous orthogonal increment process,}$$

define the mapping

$$I_Z : S\to S_Z,\ g\to I_Z(g)$$

where

$$I_Z(g)=\xi_g=\sum_{k=1}^{m}a_k\left[Z(\lambda_{k+1})-Z(\lambda_k)\right]\ \ (g=g_{\mathbf{a},\underline{\lambda}}\in S).$$

I_Z is linear:

Lemma 4.5

$$g_1,\ldots,g_N\in S,\ \mathbf{c}\in\mathbb{C}^N,\ g=\sum_{j=1}^{N}c_jg_j\Rightarrow I_Z(g)=\sum_{j=1}^{N}c_jI_Z(g_j).$$

Proof W.l.o.g. we may write

$$g_j(\lambda)=\sum_{k=1}^{m}a_{kj}1\left\{\lambda_k<\lambda\le\lambda_{k+1}\right\}\ (j=1,\ldots,N).$$

Then

$$g=\sum_{j=1}^{N}c_jg_j=\sum_{k=1}^{m}\left(\sum_{j=1}^{N}c_ja_{kj}\right)1\left\{\lambda_k<\lambda\le\lambda_{k+1}\right\}$$

so that

$$\begin{aligned}I_Z(g)&=\sum_{k=1}^{m}\left(\sum_{j=1}^{N}c_ja_{kj}\right)\left[Z(\lambda_{k+1})-Z(\lambda_k)\right]\\&=\sum_{j=1}^{N}c_j\sum_{k=1}^{m}a_{kj}\left[Z(\lambda_{k+1})-Z(\lambda_k)\right]=\sum_{j=1}^{N}c_jI_Z(g_j).\end{aligned}$$

I_Z is scalar product preserving:

Lemma 4.6

$$g_1,g_2\in S\Rightarrow\langle g_1,g_2\rangle_{L^2(F)}=\langle I_Z(g_1),I_Z(g_2)\rangle_{L^2(\Omega)}$$

Proof W.l.o.g. we may write

$$g_j(\lambda) = g_{\mathbf{a}_j,\underline{\lambda}}(\lambda) = \sum_{k=1}^{m} a_{kj} 1\{\lambda_k < \lambda \le \lambda_{k+1}\} \ (j = 1, 2).$$

Hence

$$g_1(\lambda)\overline{g_2(\lambda)} = \sum_{k=1}^{m} a_{k1}\overline{a}_{k2} 1\{\lambda_k < \lambda \le \lambda_{k+1}\},$$

$$\langle g_1, g_2\rangle = \sum_{k=1}^{m} a_{k1}\overline{a}_{k2} \int_{\lambda_k}^{\lambda_{k+1}} dF(\lambda) = \sum_{k=1}^{m} a_{k1}\overline{a}_{k2} \left[F(\lambda_{k+1}) - F(\lambda_k)\right]$$

and

$$\begin{aligned}\langle I_Z(g_1), I_Z(g_2)\rangle_{L^2(\Omega)} &= \int_{-\pi}^{\pi} \sum_{k=1}^{m} a_{k1}\overline{a}_{k2} 1\{\lambda_k < \lambda \le \lambda_{k+1}\} dF(\lambda)\\ &= \sum_{k=1}^{m} a_{k1}\overline{a}_{k2} \left[F(\lambda_{k+1}) - F(\lambda_k)\right].\end{aligned}$$

Lemma 4.7

$$\overline{S}_F = L^2(F)$$

Proof See e.g. Brockwell and Davis (1991) and Ash (1972).

Corollary 4.1 *Assumption:*

$$g_n \in S\ (n \in \mathbb{N})$$

Then

$$\exists g \in L^2(F) \text{ s.t. } g_n \underset{L^2(F)}{\to} g \Leftrightarrow \exists \xi_g \in \overline{S}_Z \text{ s.t. } \xi_{g_n} \underset{L^2(\Omega)}{\to} \xi_g$$

Proof

$$\overline{S}_F = L^2(F)$$

implies

$$g_n \in S\ (n \in \mathbb{N}) \text{ Cauchy sequence} \Leftrightarrow \exists g \in L^2(F) \text{ s.t. } g_n \underset{L^2(F)}{\to} g.$$

Due to Lemma 4.6,

$$\|g_n - g_m\| = \left\|\xi_{g_n} - \xi_{g_m}\right\|$$

so that

$$g_n \in S\ (n \in \mathbb{N}) \text{ Cauchy sequence} \Leftrightarrow \xi_{g_n} \in S_Z\ (n \in \mathbb{N}) \text{ Cauchy sequence}$$

and hence also

$$\exists g \in L^2(F) \text{ s.t. } g_n \underset{L^2(F)}{\to} g \Leftrightarrow \exists \xi_g \in \overline{S}_Z \text{ s.t. } \xi_{g_n} \underset{L^2(\Omega)}{\to} \xi_g$$

Definition 4.5 Let

$$g \in L^2(F), g_n \in S\ (n \in \mathbb{N}), g_n \underset{L^2(F)}{\to} g.$$

Then

$$I_Z(g_n) \underset{L^2(\Omega)}{\to} \xi_g \in L^2(\Omega)$$

and we define

$$I_Z(g) = \xi_g.$$

We also write

$$I_Z(g) = \int_{-\pi}^{\pi} g(\lambda)\, dZ(\lambda).$$

Definition 4.6 Let H_1, H_2 be two Hilbert spaces. A mapping $G: H_1 \to H_2$ is called an isomorphism, if it is linear, bijective and scalar product preserving.

Remark 4.2 Sometimes an isomorphism between Hilbert spaces is defined more generally as a continuous linear bijective mapping whose inverse is also continuous. The definition above is then called an isometric isomorphism. Throughout this book, "isomorphism" will be understood as an isometric isomorphism.

Lemma 4.8 *Assumptions:*

$Z(\lambda)$ $(\lambda \in [-\pi, \pi])$ *right-continuous orthogonal increment process*

$$I_Z: L^2(\Omega) \to \overline{S}_Z, g \to I_Z(g) = \xi_g$$

Then

$$I_Z = \textit{isomorphism between } L^2(\Omega) \textit{ and } \overline{S}_Z$$

Proof

1) I_Z linear:

$$g = \sum_{j=1}^{N} c_j g_j,\; g_j \in L^2(F).$$

$$g_j \in L^2(F) \Rightarrow \exists g_{nj} \in S\,(n \in \mathbb{N}) \text{ s.t. } g_n = \sum_{j=1}^{N} c_j g_{nj} \underset{L^2(F)}{\rightarrow} g\,(n \rightarrow \infty)$$

$$\Rightarrow I_Z(g) = \sum_{j=1}^{N} c_j I_Z(g_{nj}) \underset{L^2(F)}{\rightarrow} \sum_{j=1}^{N} c_j I_Z(g_j)$$

2) I_Z scalar product preserving:

$$g, h \in L^2(F) \Rightarrow \exists g_n, h_n \in S\,(n \in \mathbb{N}) \text{ s.t. } g_n \underset{L^2(F)}{\rightarrow} g,\; h_n \underset{L^2(F)}{\rightarrow} h$$

$$\Rightarrow I_Z(g_n) \underset{L^2(\Omega)}{\rightarrow} I_Z(g),\; I_Z(h_n) \underset{L^2(\Omega)}{\rightarrow} I_Z(h)$$

$$\Rightarrow \langle g, h\rangle_{L^2(F)} = \left\langle \lim_{n\to\infty} g_n, \lim_{n\to\infty} h_n \right\rangle_{L^2(F)} = \lim_{n\to\infty} \langle g_n, h_n\rangle_{L^2(F)}$$

$$= \lim_{n\to\infty} \langle I_Z(g_n), I_Z(h_n)\rangle_{L^2(\Omega)} = \left\langle \lim_{n\to\infty} I_Z(g_n), \lim_{n\to\infty} I_Z(h_n) \right\rangle_{L^2(\Omega)}$$

$$= \langle I_Z(g), I_Z(h)\rangle_{L^2(\Omega)}.$$

3) I_Z injective: (2) implies

$$\|g - h\|_{L^2(F)} = 0 \Leftrightarrow \|I_Z(g) - I_Z(h)\|_{L^2(\Omega)} = 0$$

4) I_Z surjective, i.e. $\overline{S}_Z = I_Z\left(L^2(F)\right)$:

$$g_n \in S\,(n \in \mathbb{N}) \text{ Cauchy sequence} \Leftrightarrow I_Z(g_n) \in S_Z\,(n \in \mathbb{N}) \text{ Cauchy sequence}$$

implies

$$g \in L^2(F)\ (n \in \mathbb{N}) \Leftrightarrow I_Z(g) \in \bar{S}_Z\ (n \in \mathbb{N})$$

and hence

$$\overline{S}_Z = I_Z\left(L^2(F)\right).$$

4.3.2 *Existence and Definition of Z*

4.3.2.1 Existence

Problem 4.3

$$X_t \in \mathbb{C}\ (t \in \mathbb{Z}) \text{ weakly stationary, } E(X_t) = 0,\ \gamma_X(k) = \int_{-\pi}^{\pi} e^{ik\lambda} dF_X(\lambda)$$

$$\Downarrow ?$$

$$\exists Z_X(\lambda)\ \ (\lambda \in [-\pi, \pi]) \text{ s.t.:}$$

$$Z_X = \text{orthogonal increment right-continuous process,}$$

$$X_t \underset{L^2(\Omega),a.s.}{=} \int_{-\pi}^{\pi} e^{it\lambda} dZ_X(\lambda)\ \ (t \in \mathbb{Z})$$

Solution 4.3 Define

$$e_t = \exp(it\lambda) \in L^2(F_X)\ \ (t \in \mathbb{Z}),$$

$$h_{\lambda_1,\lambda_2}(\lambda) = 1\{\lambda_1 < \lambda \le \lambda_2\} \in L^2(F_X)\ \ (\lambda_1 \le \lambda_2)$$

If

$$X_t \underset{L^2(\Omega)}{=} \int_{-\pi}^{\pi} e^{it\lambda} dZ_X(\lambda) = I_Z(e_t)\ \ (t \in \mathbb{Z}),$$

then

$$Z_X(\lambda_2) - Z_X(\lambda_1) = I_Z(h_{\lambda_1,\lambda_2}) \in L^2(\Omega).$$

(a) Abstract solution (existence of Z): apply inverse of I_Z; (b) Constructive solution: apply Corollary 3.6 with

$$h_n = \sum_{t=-n}^{n} a_t e_t \underset{L^2(F_X)}{\rightarrow} h_{\lambda_1\lambda_2}.$$

Here, an outline of the main arguments is given. For more details see Brockwell and Davis (1991).

Notation 11

$$S_{\exp} = \left\{ g \mid g(\lambda) = \sum_{j=1}^{m} a_j e_{t_j}(\lambda), \mathbf{a} \in \mathbb{C}^m, \mathbf{t} \in \mathbb{Z}^m, m \in \mathbb{N} \right\}$$

$$\overline{S}_{\exp} = L^2(F_X)\text{-closure of } S_{\exp}$$

$$S_X = \left\{ \xi \mid \xi = \sum_{j=1}^{m} a_j X_{t_j}, \mathbf{a} \in \mathbb{C}^m, \mathbf{t} \in \mathbb{Z}^m, m \in \mathbb{N} \right\}$$

$$\overline{S}_X = L^2(\Omega)\text{-closure of } S_X$$

Lemma 4.9

$$\overline{S}_{\exp} = L^2(F_X)$$

Proof See e.g. Ash (1972).

Theorem 4.1 *Assumptions:*

$$X_t \in \mathbb{C}\ (t \in \mathbb{Z}) \text{ weakly stationary, } E(X_t) = 0,\ \gamma_X(k) = \int_{-\pi}^{\pi} e^{ik\lambda} dF_X(\lambda).$$

Then

$$\exists! \text{isomorphism } \Psi_X : \overline{S}_X \rightarrow L^2(F_X) \text{ s.t.:}$$

$$\forall t \in \mathbb{Z} : \Psi_X(X_t) = e_t$$

Proof

1) Existence: Define

$$\Psi_X : S_X \rightarrow S_{\exp}, \xi = \sum_{j=1}^{k} a_j X_{t_j} \rightarrow \Psi_X(\xi) = \sum_{j=1}^{k} a_j e_{t_j}.$$

Then

$$\Psi_X(S_X) = S_{\exp} \text{ and } \Psi_X \text{ linear.}$$

For $\xi, \eta \in S_X$ we may write w.l.o.g.

$$\xi = \sum_{j=1}^{m} a_j X_{t_j}, \eta = \sum_{j=1}^{m} b_j X_{t_j}$$

so that

$$\begin{aligned}
\langle \xi, \eta \rangle_{L^2(\Omega)} &= E\left[\sum_{j=1}^{m} a_j X_{t_j} \overline{\sum_{j=1}^{m} b_j X_{t_j}}\right] \\
&= \sum_{j,l=1}^{m} a_j \overline{b_l} \gamma_X\left(t_j - t_l\right) = \int_{-\pi}^{\pi} \sum_{j,l=1}^{m} a_j \overline{b_l} e^{i(t_j - t_l)\lambda} dF_X(\lambda) \\
&= \int_{-\pi}^{\pi} \Psi_X(\xi) \overline{\Psi_X(\eta)} dF_X(\lambda) = \langle \Psi_X(\xi), \Psi_X(\eta) \rangle_{L^2(F_X)}.
\end{aligned}$$

Thus

$$\Psi_X : S_X \to S_{\exp} \text{ scalar product preserving, injective.}$$

We conclude:

$$\Psi_X : S_X \to S_{\exp} \text{ is an isomorphism,}$$

and, by extension to L^2-limits,

$$\Psi_X : \bar{S}_X \to \bar{S}_{\exp} = L^2(F_X) \text{ is an isomorphism.}$$

2) Uniqueness: Let

$$\Psi_X : \overline{S}_X \to L^2(F_X) \text{ s.t. } \forall t \in \mathbb{Z} : \Psi_X(X_t) = e_t \ (\forall t \in \mathbb{Z}),\ \Psi_X = \text{isomorphism,}$$

$$\Phi_X : \overline{S}_X \to L^2(F_X) \text{ s.t. } \forall t \in \mathbb{Z} : \Phi_X(X_t) = e_t \ (\forall t \in \mathbb{Z}),\ \Phi_X = \text{isomorphism.}$$

Then

$$\forall \xi_n \in S_X : \Psi_X(\xi_n) = \Phi_X(\xi_n),$$

and

$$\xi \in \overline{S}_X,\ \xi_n \in S_X\ (n \in \mathbb{N}),\ \lim_{n\to\infty} \|\xi - \xi_n\|_{L^2(\Omega)} = 0$$

implies

$$\begin{aligned} 0 &= \lim_{n\to\infty} \|\xi - \xi_n\|_{L^2(\Omega)} \\ &= \lim_{n\to\infty} \|\Psi_X(\xi) - \Psi_X(\xi_n)\|_{L^2(\Omega)} \\ &= \lim_{n\to\infty} \|\Phi_X(\xi) - \Phi_X(\xi_n)\|_{L^2(\Omega)} . \end{aligned}$$

Hence

$$\begin{aligned} &\|\Psi_X(\xi) - \Phi_X(\xi)\|_{L^2(F_X)} \\ &\quad \le \|\Psi_X(\xi) - \Psi_X(\xi_n)\|_{L^2(F_X)} + \|\Psi_X(\xi_n) - \Phi_X(\xi_n)\|_{L^2(F_X)} \\ &\qquad + \|\Phi_X(\xi_n) - \Phi_X(\xi)\|_{L^2(F_X)} \\ &\quad = \|\Psi_X(\xi) - \Psi_X(\xi_n)\|_{L^2(F_X)} + 0 + \|\Phi_X(\xi_n) - \Phi_X(\xi)\|_{L^2(F_X)} \end{aligned}$$

together with

$$\forall \xi \in \overline{S}_X : \exists \xi_n \in S_X\ (n \in \mathbb{N}) \text{ s.t. } \lim_{n\to\infty} \|\xi - \xi_n\|_{L^2(\Omega)} = 0$$

implies

$$\forall \xi \in \overline{S}_X : \|\Psi_X(\xi) - \Phi_X(\xi)\|_{L^2(F_X)} = 0.$$

Remark 4.3

$\Psi_X : \overline{S}_X \to L^2(F)$ isomorphism $\Rightarrow \exists \Psi_X^{-1} : L^2(F) \to \overline{S}_X$ and $\Psi_X^{-1} =$ isomorphism

Theorem 4.2 *Assumptions:*

$$X_t \in \mathbb{C}\ (t \in \mathbb{Z}) \text{ weakly stationary, } E(X_t) = 0,\ \gamma_X(k) = \int_{-\pi}^{\pi} e^{ik\lambda} dF_X(\lambda),$$

$$h_{\lambda_1,\lambda_2}(\nu) = 1\{\lambda_1 < \nu \le \lambda\}\ (\lambda_2 \ge \lambda_1).$$

Define

$$Z_X(\lambda) = \Psi_X^{-1}(h_{-\pi,\lambda})\ (\lambda \in (-\pi, \pi]),\ Z_X(-\pi) = 0.$$

Then

$$Z_X = \textit{right-continuous orthogonal increment process on } [-\pi, \pi],$$

and

$$E\left[|Z_X(\lambda_2) - Z_X(\lambda_1)|^2\right] = F_X(\lambda_2) - F_X(\lambda_1) \ (\lambda_2 \geq \lambda_1).$$

Proof

1) $E(Z_X)$:

$$Z_X(-\pi) = 0 \Rightarrow E[Z_X(-\pi)] = 0.$$

For $-\pi < \lambda \leq \pi$,

$$h_{-\pi,\lambda} \in L^2(F_X), \Psi_X \text{ bijective} \Rightarrow \Psi_X^{-1}(h_{-\pi,\lambda}) \in \overline{S}_X$$

$$\Rightarrow \exists \xi_n \in S_X \ (n \in \mathbb{N}) \text{ s.t. } \lim_{n\to\infty} \left\| \Psi_X^{-1}(h_{-\pi,\lambda}) - \xi_n \right\|^2_{L^2(\Omega)} = 0.$$

Hence,

$$E(\xi_n) = \langle \xi_n, 1 \rangle_{L^2(\Omega)} = 0 \ (\xi_n \in S_X)$$

implies

$$\begin{aligned} E[Z_X(\lambda)] &= E\left[\Psi_X^{-1}(h_{-\pi,\lambda})\right] = \left\langle \Psi_X^{-1}(h_{-\pi,\lambda}), 1 \right\rangle_{L^2(\Omega)} \\ &= \lim_{n\to\infty} \langle \xi_n, 1 \rangle_{L^2(\Omega)} = 0 \end{aligned}$$

2) $var(Z_x)$:

$$\Psi_X \text{ isomorphism} \Rightarrow \Psi_X^{-1} \text{ isomorphism}$$

Hence,

$$\begin{aligned} \|Z_X(\lambda_2) - Z_X(\lambda_1)\|^2_{L^2(\Omega)} &= \left\| \Psi_X^{-1}(h_{-\pi,\lambda_2}) - \Psi_X^{-1}(h_{-\pi,\lambda_1}) \right\|^2_{L^2(\Omega)} \\ &= \|h_{-\pi,\lambda_2} - h_{-\pi,\lambda_1}\|^2_{L^2(F_X)} = \int_{\lambda_1}^{\lambda_2} dF_X(\lambda) \\ &= F_X(\lambda_2) - F_X(\lambda_1) \end{aligned}$$

3) right-continuous:

$$F_X \text{ right-continuous, 2)} \Rightarrow Z_X \text{ right-continuous process}$$

4) orthogonal increments: For $-\pi \le \lambda_1 < \lambda_2 \le \lambda_3 < \lambda_4 \le \pi$,

$$cov\left(Z_X(\lambda_4) - Z_X(\lambda_3), Z_X(\lambda_2) - Z_X(\lambda_1)\right) = \left\langle \Psi_X^{-1}\left(h_{\lambda_3,\lambda_4}\right), \Psi_X^{-1}\left(h_{\lambda_1,\lambda_2}\right)\right\rangle_{L^2(\Omega)}$$

$$= \langle h_{\lambda_3,\lambda_4}, h_{\lambda_1,\lambda_2}\rangle_{L^2(F)} = \int_{-\pi}^{\pi} h_{\lambda_3,\lambda_4}(\lambda)\, h_{\lambda_1,\lambda_2}(\lambda)\, dF_X(\lambda) = 0.$$

Lemma 4.10 *Assumptions:*

$$X_t \in \mathbb{C}\ (t \in \mathbb{Z}) \text{ weakly stationary, } E(X_t) = 0,\ \gamma_X(k) = \int_{-\pi}^{\pi} e^{ik\lambda} dF_X(\lambda),$$

$$\Psi_X : \overline{S}_X \to L^2(F_X) \text{ isomorphism s.t. } \forall t \in \mathbb{Z} : \Psi(X_t) = e_t,$$

$$h_{-\lambda_1,\lambda_2}(\nu) = 1\{\lambda_1 < \nu \le \lambda\}\ (\lambda_2 \ge \lambda_1),$$

$$Z_X(\lambda) = \Psi_X^{-1}(h_{-\pi,\lambda})\ (\lambda \in (-\pi, \pi]),\ Z_X(-\pi) = 0.$$

Then

$$\forall g \in L^2(F_X) : \Psi_X^{-1}(g) \underset{a.s.}{=} I_{Z_X}(g).$$

Proof Let

$$g = g_{\mathbf{a},\underline{\lambda}} = \sum_{k=1}^{m} a_k \left(h_{-\pi,\lambda_{k+1}} - h_{-\pi,\lambda_k}\right) \in S.$$

Since

$$Z_X(\lambda) = \Psi_X^{-1}(h_{-\pi,\lambda}),$$

$$\Psi_X : \overline{S}_X \to L^2(F_X) \text{ isomorphism},$$

$$I_{Z_X} : L^2(F_X) \to \overline{S}_X \text{ isomorphism}$$

we obtain

$$I_{Z_X}(g) = \sum_{k=1}^{m} a_k \left[Z_X(\lambda_{k+1}) - Z_X(\lambda_k)\right] = \sum_{j=1}^{k} a_k \left[\Psi_X^{-1}\left(h_{-\pi,\lambda_{k+1}}\right) - \Psi_X^{-1}\left(h_{-\pi,\lambda_k}\right)\right]$$

$$= \Psi_X^{-1}\left(\sum_{j=1}^{k} a_j \left(h_{-\pi,\lambda_{j+1}} - h_{-\pi,\lambda_j}\right)\right) = \Psi_X^{-1}(g).$$

Thus,

$$\forall g \in S : I_{Z_X}(g) = \Psi_X^{-1}(g).$$

Now

$$g \in L^2(F_X) \Rightarrow \exists g_n \in S\ (n \in \mathbb{N}) \text{ s.t. } \|g - g_n\|_{L^2(F)} \to 0$$

so that

$$\begin{aligned}
\left\| I_{Z_X}(g) - \Psi_X^{-1}(g) \right\| &\leq \lim_{n\to\infty} \| I_{Z_X}(g) - I_{Z_X}(g_n) \| \\
&+ \lim_{n\to\infty} \left\| I_{Z_X}(g_n) - \Psi_X^{-1}(g_n) \right\| \\
&+ \lim_{n\to\infty} \left\| \Psi_X^{-1}(g_n) - \Psi_X^{-1}(g) \right\| \\
&= 0.
\end{aligned}$$

Theorem 4.3 *Assumptions:*

$$X_t \in \mathbb{C}\ (t \in \mathbb{Z}) \text{ weakly stationary, } E(X_t) = 0,\ \gamma_X(k) = \int_{-\pi}^{\pi} e^{ik\lambda} dF_X(\lambda)$$

Then

$$\exists! Z_X \text{ right-continuous orthogonal increment process on } [-\pi, \pi] \text{ s.t.:}$$

$$Z_X(-\pi) = 0,$$

for $-\pi \leq \lambda_1 \leq \lambda_2 \leq \pi$,

$$E\left[|Z_X(\lambda_2) - Z_X(\lambda_1)|^2 \right] = F_X(\lambda_2) - F_X(\lambda_1)$$

and

$$X_t \underset{a.s.}{=} I_{Z_X}(e_t) = \int_{-\pi}^{\pi} e^{it\lambda} dZ_X(\lambda)\ (t \in \mathbb{Z}).$$

Proof

1) Existence: Define

$$Z_X(\lambda) = \Psi_X^{-1}(h_{-\pi,\lambda})\ (\lambda \in (-\pi, \pi]),\ Z_X(-\pi) = 0,\ Z_X(\lambda) = Z_X(\pi)\ (\lambda > \pi)$$

Theorem 4.2 and Lemma 4.10 imply

$$Z_X = \text{right-continuous orthogonal increment process,}$$

$$E\left[|Z_X(\lambda_2) - Z_X(\lambda_1)|^2\right] = F_X(\lambda_2) - F_X(\lambda_1),$$

$$\forall g \in L^2(F_X) : \Psi_X^{-1}(g) = I_{Z_X}(g).$$

In particular,

$$\forall t \in \mathbb{Z} : \Psi_X^{-1}(e_t) = I_{Z_X}(e_t).$$

Now, by definition of Ψ_X and I_{Z_X},

$$\Psi_X^{-1}(e_t) = X_t \text{ and } I_{Z_X}(e_t) = \int_{-\pi}^{\pi} e^{it\lambda} dZ_X(\lambda)$$

so that

$$\forall t \in \mathbb{Z} : X_t = I_{Z_X}(e_t) = \int_{-\pi}^{\pi} e^{it\lambda} dZ_X(\lambda).$$

2) Uniqueness: Let

$$W \text{ right-continuous, orthogonal increment process s.t:}$$

$$W(-\pi) = 0,$$

for $-\pi \le \lambda_1 \le \lambda_2 \le \pi$,

$$E\left[|W(\lambda_2) - W(\lambda_1)|^2\right] = F_X(\lambda_2) - F_X(\lambda_1)$$

and

$$X_t = I_W(e_t) \ \ (t \in \mathbb{Z}).$$

Then

$$\int_{-\pi}^{\pi} e^{it\lambda} dZ_X(\lambda) \underset{a.s.}{=} \int_{-\pi}^{\pi} e^{it\lambda} dW(\lambda) \ \ (t \in \mathbb{Z})$$

and

$$h_{-\pi,\lambda} \in L^2(F_X) \cap L^2(W),$$

imply

$$Z_X(\lambda) = \int h_{-\pi,\lambda}(\nu)\, dZ_X(\nu) \underset{a.s.}{=} \int h_{-\pi,\lambda}(\nu)\, dW(\nu) = W(\lambda) \;\; (\lambda \in [-\pi, \pi]).$$

Definition 4.7 Let

$$X_t \in \mathbb{C}\ (t \in \mathbb{Z}) \text{ weakly stationary, } E(X_t) = 0,\ \gamma_X(k) = \int_{-\pi}^{\pi} e^{ik\lambda} dF_X(\lambda),$$

$$Z_X \text{ right-continuous orthogonal increment process,}$$

$$Z_X(-\pi) = 0,$$

$$\forall t \in \mathbb{Z} : X_t \underset{a.s.}{=} \int_{-\pi}^{\pi} e^{it\lambda} dZ_X(\lambda).$$

Then Z_X is called spectral measure of X_t $(t \in \mathbb{Z})$.

4.3.2.2 Calculation of Z

Problem 4.4 Let

$$X_t \in \mathbb{C}\ (t \in \mathbb{Z}) \text{ weakly stationary, } E(X_t) = 0,\ \gamma_X(k) = \int_{-\pi}^{\pi} e^{ik\lambda} dF_X(\lambda).$$

Calculate the spectral measure $Z_X(\lambda)$.

Solution 4.4

$$h_{\lambda_1,\lambda_2}(\lambda) = 1\{\lambda_1 < \lambda \le \lambda_2\},\ e_t(\lambda) = e^{it\lambda},\ a_t = \frac{1}{2\pi}\int_{-\pi}^{\pi} e^{-it\lambda} h_{\lambda_1,\lambda_2}(\lambda)\, d\lambda$$

For $\lambda_1, \lambda_2 \in J_{cont}(F_X)$, Corollary 3.6 implies,

$$h_n = \sum_{t=-n}^{n} a_t e_t \underset{L^2(F_X)}{\rightarrow} h_{\lambda_1,\lambda_2}.$$

We will show

$$I_{Z_X}(h_{\lambda_1,\lambda_2}) = \sum_{t=-\infty}^{\infty} a_t X_t,$$

which then implies

$$Z_X(\lambda_2) - Z_X(\lambda_1) = \sum_{t=-\infty}^{\infty} a_t X_t.$$

Theorem 4.4 *Assumptions:*

$$X_t \in \mathbb{C}\ (t \in \mathbb{Z})\ \text{weakly stationary},\ E(X_t) = 0,$$

$$Z_X = \text{spectral measure of } X_t,$$

$$\lambda_2 \geq \lambda_1,\ \lambda_1, \lambda_2 \in J_{cont}(F_X),$$

$$a_t = \frac{1}{2\pi} \int_{-\pi}^{\pi} e^{-it\lambda} h_{\lambda_1,\lambda_2}(\lambda)\, d\lambda.$$

Then

$$Z_X(\lambda_2) - Z_X(\lambda_1) \underset{a.s.}{=} \sum_{t=-\infty}^{\infty} a_t X_t.$$

Proof

$$\lambda_1, \lambda_2 \in J_{cont}(F_X)$$

implies

$$h_n = \sum_{t=-n}^{n} a_t e_t \underset{L^2(F_X)}{\rightarrow} h_{\lambda_1,\lambda_2}$$

and hence

$$\Psi_X^{-1}(h_n) \underset{L^2(\Omega)}{\rightarrow} \Psi_X^{-1}(h_{\lambda_1,\lambda_2}) = I_{Z_X}(h_{\lambda_1,\lambda_2}) = Z_X(\lambda_2) - Z_X(\lambda_1).$$

Since

$$\Psi_X^{-1} = \text{isomorphism},$$

we obtain

$$\Psi_X^{-1}(h_n) = \Psi_X^{-1}\left(\sum_{t=-n}^{n} a_t e_t\right) = \sum_{t=-n}^{n} a_t \Psi_X^{-1}(e_t) = \sum_{t=-n}^{n} a_t X_t$$

so that

$$\sum_{t=-\infty}^{\infty} a_t X_t \underset{a.s.}{=} Z_X(\lambda_2) - Z_X(\lambda_1).$$

4.3.3 *Interpretation of the Spectral Representation*

$$X_t = \int_{-\pi}^{\pi} e^{it\lambda} dZ_X(\lambda) = \text{orthogonal decomposition}$$

$$e^{it\lambda} dZ_X(\lambda) \ (\lambda \in [-\pi, \pi]) = \text{periodic components}$$

$$T = \frac{2\pi}{\lambda} = \text{period corresponding to frequency } \lambda$$

$$dZ_X(\lambda) = \text{random amplitude at frequency } \lambda$$

$$\begin{aligned} F_X(\lambda + \delta) - F_X(\lambda) &= E\left(|Z_X(\lambda + \delta) - Z_X(\lambda)|^2\right) \\ &= var\,(Z_X(\lambda + \delta) - Z_X(\lambda)) \\ &= \text{average squared amplitude for } \nu \in (\lambda, \lambda + \delta] \end{aligned}$$

Notation 12 *Symbolic infinitesimal notation*

$$dF_X(\lambda) = E\left(|dZ_X(\lambda)|^2\right) = var\,(dZ_X(\lambda))$$

$$var\,(X_t) = \gamma_X(0) = \int_{-\pi}^{\pi} dF_X(\lambda) = \int_{-\pi}^{\pi} E\left(|dZ_X(\lambda)|^2\right)$$

4.4 Further Properties

4.4.1 *Relationship Between* Re Z *and* Im Z

Lemma 4.11 *Assumptions:*

$$Z(\lambda) = \textit{spectral measure},$$

$$Z_1(\lambda) = \operatorname{Re} Z(\lambda),\, Z_2(\lambda) = \operatorname{Im} Z(\lambda),$$

$$-\pi \le \lambda_1 \le \lambda_2 \le \lambda_3 \le \lambda_4 \le \pi.$$

Then

$$E\left[\left(Z_1(\lambda_2)-Z_1(\lambda_1)\right)\left(Z_1(\lambda_4)-Z_1(\lambda_3)\right)\right]$$
$$=-E\left[\left(Z_2(\lambda_2)-Z_2(\lambda_1)\right)\left(Z_2(\lambda_4)-Z_2(\lambda_3)\right)\right]$$

and

$$E\left[\left(Z_2(\lambda_2)-Z_2(\lambda_1)\right)\left(Z_1(\lambda_4)-Z_1(\lambda_3)\right)\right]$$
$$=E\left[\left(Z_1(\lambda_2)-Z_1(\lambda_1)\right)\left(Z_2(\lambda_4)-Z_2(\lambda_3)\right)\right].$$

Proof

$$E\left[\left(Z(\lambda_2)-Z(\lambda_1)\right)\overline{\left(Z(\lambda_4)-Z(\lambda_3)\right)}\right]=0$$

implies

$$0=\operatorname{Re}E\left[\left(Z(\lambda_2)-Z(\lambda_1)\right)\overline{\left(Z(\lambda_4)-Z(\lambda_3)\right)}\right]$$
$$=E\left[\left(Z_1(\lambda_2)-Z_1(\lambda_1)\right)\left(Z_1(\lambda_4)-Z_1(\lambda_3)\right)\right]$$
$$+E\left[\left(Z_2(\lambda_2)-Z_2(\lambda_1)\right)\left(Z_2(\lambda_4)-Z_2(\lambda_3)\right)\right]$$

and

$$0=\operatorname{Im}E\left[\left(Z(\lambda_2)-Z(\lambda_1)\right)\overline{\left(Z(\lambda_4)-Z(\lambda_3)\right)}\right]$$
$$=E\left[\left(Z_2(\lambda_2)-Z_2(\lambda_1)\right)\left(Z_1(\lambda_4)-Z_1(\lambda_3)\right)\right]$$
$$-E\left[\left(Z_1(\lambda_2)-Z_1(\lambda_1)\right)\left(Z_2(\lambda_4)-Z_2(\lambda_3)\right)\right].$$

4.4.2 Frequency

$$X_t=\int_{-\pi}^{\pi}e^{it\lambda}dZ_X(\lambda)\ \ (t\in\mathbb{Z})$$

$$\exp(it\lambda)=\exp(i(t+s)\lambda)\ \ (t\in\mathbb{Z})\Leftrightarrow s=\frac{2\pi}{\lambda}j\ (j\in\mathbb{Z})$$

$$\lambda\in(0,\pi]\Rightarrow \text{period } T=\frac{2\pi}{\lambda}$$

Instead of $\lambda = 2\pi/T$ one sometimes uses $\nu = 1/T$:

$$\nu = \frac{\lambda}{2\pi} \in (0, \frac{1}{2}] \Rightarrow \text{period } T = \frac{1}{\nu}$$

$$X_t = \int_{-\pi}^{\pi} e^{it\lambda} dZ_X(\lambda) = \int_{-1/2}^{1/2} e^{i2\pi t\nu} d\tilde{Z}_X(\nu),$$

$$\gamma_X(k) = \int_{-1/2}^{1/2} e^{i2\pi k\nu} d\tilde{F}_X(\nu).$$

4.4.3 Overtones

Definition 4.8 Let

$$0 < T < \infty,$$

$$g(t) = \sum_{j=0}^{\infty} a_j \cos\left(\frac{2\pi j}{T}t\right) + \sum_{j=1}^{\infty} b_j \sin\left(\frac{2\pi j}{T}t\right) \in \mathbb{R}\ (t \in \mathbb{R})$$

and

$$\max\{|a_1|, |b_1|\} > 0.$$

Then, for $j \geq 1$,

$$\lambda_j = \frac{2\pi}{T}j = j\text{th harmonic or partial of } g.$$

An alternative terminology is

$$\lambda_0^* = \lambda_1 = \text{fundamental frequency or fundamental tone}$$

and

$$\lambda_j^* = \lambda_{j+1} = j\text{th overtone } (j \geq 1).$$

Remark 4.4 The definition of overtones depends on the chosen orthogonal basis in $L^2[-T/2, T/2]$. In the definition above, the basis is

$$\cos\left(\frac{2\pi j}{T}t\right)\ (j = 0, 1, 2, \ldots),\ \sin\left(\frac{2\pi j}{T}t\right)\ (j = 1, 2, \ldots).$$

Remark 4.5 Deterministic periodic functions with a finite number of overtones can be understood as sample paths of a harmonic process.

A similar phenomenon can be observed for some purely stochastic weakly stationary processes with an existing spectral density f_X, in the sense that f_X has local maxima at multiples of a fundamental frequency.

Example 4.1 Let

$$\varepsilon_t \, (t \in \mathbb{Z}) \text{ iid}, E(\varepsilon_t) = 0, \sigma_\varepsilon^2 = var(\varepsilon_t) < \infty, \psi > 0,$$

$$X_t = \varepsilon_t + \psi \varepsilon_{t-12} \, (t \in \mathbb{Z}).$$

Then

$$f_X(\lambda) = \frac{\sigma_\varepsilon^2}{2\pi} \left(1 + \psi^2 + 2\psi \cos 12\lambda\right)$$

has local maxima at

$$\lambda_j = \frac{2\pi}{12} j \, (j = 1, \ldots, 6).$$

4.4.4 Why Are Frequencies Restricted to the Range $[-\pi, \pi]$?

Theorem 4.5 *Assumptions:*

$$X_t \in \mathbb{R} \, (t \in \mathbb{R}) \text{ weakly stationary, } E(X_t) = 0.$$

Then

$$\exists! F_X \text{ gcdf on } \mathbb{R} \text{ s.t.:}$$

$$\gamma_X(u) = \int_{-\infty}^{\infty} e^{iu\lambda} dF_X(\lambda) \ \ (u \in \mathbb{R}).$$

Moreover,

$$\exists! Z_X \text{ right-continuous orthogonal increment process on } \mathbb{R} \text{ s.t.:}$$

$$Z_X(-\infty) = 0,$$

$$E\left[|Z_X(\lambda_2) - Z_X(\lambda_1)|^2\right] = F_X(\lambda_2) - F_X(\lambda_1) \ \ (\lambda_2 \geq \lambda_1),$$

$$X_t \underset{a.s.}{=} \int_{-\infty}^{\infty} e^{it\lambda} dZ_X(\lambda) \ (t \in \mathbb{R}).$$

Proof Similar to Theorem 4.3.

Problem 4.5 (Discrete Sampling)

$$X_t = \int_{-\infty}^{\infty} e^{it\lambda} dZ_X(\lambda) \ (t \in \mathbb{R}),$$

$$0 < \Delta t < \infty, \ t_m = m \cdot \Delta t, \ m \in \mathbb{Z},$$

$$Y_m = X_{t_m} = \text{observed process } (m \in \mathbb{Z})$$

$$\Rightarrow \text{spectral representation of } Y_m?$$

Solution 4.5 Follows from Theorem 4.5 and periodicity of sin and cos.

Definition 4.9

$$\lambda_{\text{sampling}} = \frac{2\pi}{\Delta t} = \text{sampling rate},$$

$$\lambda_{\text{Nyquist}} = \frac{\lambda_{\text{sampling}}}{2} = \frac{\pi}{\Delta t} = \text{Nyquist frequency}$$

or, when frequency is expressed as the inverse of the period,

$$\nu_{\text{sampling}} = \frac{1}{\Delta t} = \text{sampling rate},$$

$$\nu_{\text{Nyquist}} = \frac{\nu_{\text{sampling}}}{2} = \frac{1}{2}\frac{1}{\Delta t} = \text{Nyquist frequency}$$

(see e.g. Nyquist 1928; Shannon 1948).

Theorem 4.6 *Assumptions:*

$$X_t \underset{a.s.}{=} \int_{-\infty}^{\infty} e^{it\lambda} dZ_X(\lambda) \ (t \in \mathbb{R}),$$

$$0 < \Delta t < \infty, \ t_m = m \cdot \Delta t, \ m \in \mathbb{Z},$$

$$Y_m = X_{t_m} \ (m \in \mathbb{Z})$$

Then

$$Y_m \ (m \in \mathbb{Z}) \text{ weakly stationary, } E(Y_m) = 0,$$

$$Y_m \underset{a.s.}{=} \int_{-\lambda_{Nyquist}}^{\lambda_{Nyquist}} \exp(i \cdot m \cdot \Delta t \cdot \lambda)\, dZ_Y(\lambda) \ \ (m \in \mathbb{Z}),$$

$$\gamma_Y(k) = \int_{-\lambda_{Nyquist}}^{\lambda_{Nyquist}} \exp(ik \cdot \Delta t \cdot \lambda)\, dF_Y(\lambda) \ \ (k \in \mathbb{Z}),$$

where, for $-\lambda_{Nyquist} \le \lambda_1 \le \lambda_2 \le \lambda_{Nyquist}$,

$$\begin{aligned} Z_Y(\lambda_2) - Z_Y(\lambda_1) &= \sum_{j=-\infty}^{\infty} \left[Z_X\left(\lambda_2 + 2j \cdot \lambda_{Nyquist}\right) - Z_X\left(\lambda_1 + 2j \cdot \lambda_{Nyquist}\right)\right] \\ &= \sum_{j=-\infty}^{\infty} \left[Z_X\left(\lambda_2 + j \cdot \lambda_{sampling}\right) - Z_X\left(\lambda_1 + j \cdot \lambda_{sampling}\right)\right], \end{aligned}$$

$$\begin{aligned} F_Y(\lambda_2) - F_Y(\lambda_1) &= \sum_{j=-\infty}^{\infty} \left[F_X\left(\lambda_2 + 2j \cdot \lambda_{Nyquist}\right) - F_X\left(\lambda_1 + 2j \cdot \lambda_{Nyquist}\right)\right] \\ &= \sum_{j=-\infty}^{\infty} \left[F_X\left(\lambda_2 + j \cdot \lambda_{sampling}\right) - F_X\left(\lambda_1 + j \cdot \lambda_{sampling}\right)\right]. \end{aligned}$$

Proof

$$\exp(i \cdot m \cdot \Delta t \cdot \lambda) = \exp\left\{ i \cdot m \cdot \Delta t \cdot \left(\lambda + \frac{2\pi}{\Delta t} j\right)\right\} \ \ (j \in \mathbb{Z})$$

implies

$$\begin{aligned} Y_m &= \int_{-\infty}^{\infty} \exp(i \cdot m \cdot \Delta t \cdot \lambda)\, dZ_X(\lambda) \\ &= \sum_{j=-\infty}^{\infty} \int_{(j-\frac{1}{2})\frac{2\pi}{\Delta t}}^{(j+\frac{1}{2})\frac{2\pi}{\Delta t}} \exp(i \cdot m \cdot \Delta t \cdot \lambda)\, dZ_X(\lambda) \\ &= \int_{-\frac{\pi}{\Delta t}}^{\frac{\pi}{\Delta t}} \exp(i \cdot m \cdot \Delta t \cdot \lambda) \sum_{j=-\infty}^{\infty} dZ_X\left(\lambda + \frac{2\pi}{\Delta t} j\right) \\ &= \int_{-\lambda_{\text{Nyquist}}}^{\lambda_{\text{Nyquist}}} \exp(i \cdot m \cdot \Delta t \cdot \lambda)\, dZ_Y(\lambda) \end{aligned}$$

Corollary 4.2 *Assumptions:*

$$X_t \in \mathbb{R}\ (t \in \mathbb{R})\ \text{weakly stationary},\ E(X_t) = 0,\ \exists f_X = F_X',$$

$$0 < \Delta t < \infty,\ t_m = m \cdot \Delta t,\ m \in \mathbb{Z},$$

$$Y_m = X_{t_m}\ (m \in \mathbb{Z})$$

Then

$$f_Y(\lambda) = \sum_{k=-\infty}^{\infty} f_X\left(\lambda + 2j \cdot \lambda_{Nyquist}\right)\ \left(\lambda \in \left[-\lambda_{Nyquist}, \lambda_{Nyquist}\right]\right).$$

Conclusion 8 *Corollary 4.2 means that values of f_X outside $[-\lambda_{Nyquist}, \lambda_{Nyquist}]$ are added to values of f_X in $[-\lambda_{Nyquist}, \lambda_{Nyquist}]$. In particular, if f_X is large or has relevant local maxima for $\lambda > \lambda_{Nyquist}$, discrete sampling leads to incorrectly identifying high-frequency oscillations with lower frequency periodicities. This effect is called aliasing.*

Various solutions have been proposed to alleviate the problem. Some of the solutions are:

1. Increased sampling rate:

$$\lim_{\Delta t \to 0} \lambda_{\text{Nyquist}} = \lim_{\Delta t \to 0} \frac{\pi}{\Delta t} = \infty.$$

   To avoid aliasing up to frequency

$$\lambda_{\max} = \frac{2\pi}{T_{\min}},$$

   choose Δt such that

$$\lambda_{\text{Nyquist}} > \lambda_{\max} \Leftrightarrow \Delta t < \frac{1}{2} T_{\min}.$$

2. Before sampling, filter out frequencies $\lambda \geq \lambda_{\max}$.
3. Observe at random time points.

Example 4.2 The commonly stated range of human hearing is 20 Hz–20 kHz. In terms of frequencies $\lambda \in [-\pi, \pi]$ or $\nu \in [-\frac{1}{2}, \frac{1}{2}]$, this means

$$\lambda_{\max} = 2\pi \cdot 20{,}000\,\text{Hz},\ \nu_{\max} = 20{,}000\,\text{Hz}$$

$$T_{\min} = \frac{1}{20{,}000}\ \text{s}.$$

A minimal sampling rate for digital recordings is therefore given by

$$\Delta t < \frac{1}{2}T_{\min} = \frac{1}{40{,}000}\,\mathrm{s} \Leftrightarrow \nu_{\text{sampling}} = \frac{1}{\Delta t} > 40{,}000\,\mathrm{Hz}$$

For CD recordings one generally uses $\nu_{\text{sampling}} = 44{,}100\,\mathrm{Hz}$.

4.5 Linear Filters and the Spectral Representation

4.5.1 *Effect on the Spectral Representation*

Theorem 4.7 *Assumptions:*

$$X_t \underset{a.s.}{=} \int_{-\pi}^{\pi} e^{it\lambda} dZ_X(\lambda) \in \mathbb{R} \ (t \in \mathbb{Z}),$$

$$A_n\left(e^{-i\lambda}\right) = \sum_{j=-n}^{n} a_j e^{-ij\lambda} \underset{L^2(F_X)}{\rightarrow} A\left(e^{-i\lambda}\right) = \sum_{j=-\infty}^{\infty} a_j e^{-ij\lambda}$$

$$Y_{t,n} = A_n(B) X_t \ (t \in \mathbb{Z}).$$

Then

$$\forall t \in \mathbb{Z} : \exists Y_t \in L^2(\Omega) \text{ s.t.:}$$

$$Y_{t,n} \underset{L^2(\Omega)}{\rightarrow} Y_t,$$

$$Y_t = \int_{-\pi}^{\pi} e^{it\lambda} A\left(e^{-i\lambda}\right) dZ_X(\lambda) \ (a.s. \text{ and } L^2(\Omega)),$$

$$\gamma_Y(k) = \int_{-\pi}^{\pi} e^{ik\lambda} \left|A\left(e^{-i\lambda}\right)\right|^2 dF_X(\lambda) \ (k \in \mathbb{Z}).$$

We then write

$$Y_t = \int_{-\pi}^{\pi} e^{it\lambda} dZ_Y(\lambda), \ \gamma_Y(k) = \int_{-\pi}^{\pi} e^{ik\lambda} dF_Y(\lambda)$$

with

$$dZ_Y(\lambda) = A\left(e^{-i\lambda}\right) dZ_X(\lambda), \ dF_Y(\lambda) = \left|A\left(e^{-i\lambda}\right)\right|^2 dF_X(\lambda).$$

Proof Using the notation

$$A_n(z) = \sum_{j=-n}^{n} a_j z^j \ (z \in \mathbb{C}),$$

we have

$$Y_{t,n} = \sum_{j=-n}^{n} a_j X_{t-j} = \sum_{j=-n}^{n} a_j \int_{-\pi}^{\pi} e^{i(t-j)\lambda} dZ_X(\lambda) = \int_{-\pi}^{\pi} e^{it\lambda} A_n\left(e^{-i\lambda}\right) dZ_X(\lambda)$$

Then

$$A_n \underset{L^2(F_X)}{\rightarrow} A,$$

$$\|Y_{t,n} - Y_{t,m}\|^2_{L^2(\Omega)} = \|A_n - A_m\|^2_{L^2(F_X)},$$

and

$$L^2(\Omega) \text{ complete},$$

imply

$$\exists Y_t \in L^2(\Omega) \text{ s.t. } Y_{t,n} \underset{L^2(\Omega)}{\rightarrow} Y_t = \int_{-\pi}^{\pi} e^{it\lambda} A\left(e^{-i\lambda}\right) dZ_X(\lambda).$$

Example 4.3

$$\varepsilon_t \in \mathbb{R}, E(\varepsilon_t) = 0, cov(\varepsilon_s, \varepsilon_t) = \sigma_\varepsilon^2 \delta_{st} < \infty$$

$$\psi(B) = 1 + \sum_{j=1}^{q} \psi_j B^j$$

$$X_t = \psi(B)\varepsilon_t$$

Then

$$f_\varepsilon(\lambda) = \frac{\sigma_\varepsilon^2}{2\pi} \ (\lambda \in [-\pi, \pi]),$$

$$A\left(e^{-i\lambda}\right) = \psi\left(e^{-i\lambda}\right) = 1 + \sum_{j=1}^{q} \psi_j e^{-ij\lambda},$$

$$f_X(\lambda) = \frac{\sigma_\varepsilon^2}{2\pi} \left|\psi\left(e^{-i\lambda}\right)\right|^2 = \frac{\sigma_\varepsilon^2}{2\pi} \left|1 + \sum_{j=1}^{q} \psi_j e^{-ij\lambda}\right|^2$$

Theorem 4.8 (Inversion of Filters) *Assumptions:*

$$X_t = \int_{-\pi}^{\pi} e^{it\lambda} dZ_X(\lambda) \ (t \in \mathbb{Z})$$

$$A_n\left(e^{-i\lambda}\right) = \sum_{j=-n}^{n} a_j e^{-ij\lambda} \underset{L^2(F_X)}{\rightarrow} A\left(e^{-i\lambda}\right) = \sum_{j=-\infty}^{\infty} a_j e^{-ij\lambda}$$

$$Y_t = A(B) X_t = \int_{-\pi}^{\pi} e^{it\lambda} dZ_Y(\lambda)$$

$$\exists \Lambda \subset [-\pi, \pi] \text{ s.t. } \int_{\Lambda} dF_X(\lambda) = 0 \text{ and } A\left(e^{-i\lambda}\right) \neq 0 \ (\lambda \notin \Lambda).$$

Then

$$X_t = \int_{-\pi}^{\pi} e^{it\lambda} \frac{1}{A\left(e^{-i\lambda}\right)} dZ_Y(\lambda)$$

and

$$dF_X(\lambda) = \left|A\left(e^{-i\lambda}\right)\right|^{-2} dF_Y(\lambda).$$

Proof

$$\forall \lambda \notin \Lambda : A^{-1}\left(e^{-i\lambda}\right) A\left(e^{-i\lambda}\right) = 1$$

Hence,

$$\forall t \in \mathbb{Z} : \exists U_t = \int_{-\pi}^{\pi} e^{it\lambda} A^{-1}\left(e^{-i\lambda}\right) dZ_Y(\lambda) \in \mathbb{R}$$

and

$$\|X_t - U_t\|_{L^2(\Omega)}^2 = \int_{-\pi}^{\pi} \left|1 - A^{-1}\left(e^{-i\lambda}\right) A\left(e^{-i\lambda}\right)\right|^2 dF_X(\lambda) = 0.$$

Lemma 4.12 *Assumption:*

$$\sum_{j=-\infty}^{\infty} |a_j| < \infty$$

Then

$$\forall \textit{spectral distributions } F_X :$$

$$A_n\left(e^{-i\lambda}\right) \underset{L^2(F_X)}{\rightarrow} A\left(e^{-i\lambda}\right).$$

Proof

$$\sum_{j=-\infty}^{\infty} \left|a_j\right| < \infty \Rightarrow \lim_{n\to\infty} \sup_{\lambda\in[-\pi,\pi]} \left|A_n\left(e^{-i\lambda}\right) - A\left(e^{-i\lambda}\right)\right| = 0$$

Hence,

$$\|A_n - A\|^2_{L^2(F_X)} \le F(\pi) \sup_{\lambda\in[-\pi,\pi]} \left|A_n\left(e^{-i\lambda}\right) - A\left(e^{-i\lambda}\right)\right|^2 \to 0$$

Lemma 4.13 *Assumption:*

$$\sum_{j=-\infty}^{\infty} a_j^2 < \infty$$

$$\exists f_X = F_X', f_X(\lambda) \le M < \infty \; (\lambda \in [-\pi,\pi])$$

Then

$$A_n\left(e^{-i\lambda}\right) \underset{L^2(F_X)}{\rightarrow} A\left(e^{-i\lambda}\right).$$

Proof

$$\sum_{j=-\infty}^{\infty} a_j^2 < \infty \Rightarrow A_n\left(e^{-i\lambda}\right) \underset{L^2[-\pi,\pi]}{\rightarrow} A\left(e^{-i\lambda}\right)$$

Hence,

$$\begin{aligned}
\lim_{n\to\infty} \|A_n - A\|^2_{L^2(F_X)} &= \lim_{n\to\infty} \int \left|A_n\left(e^{-i\lambda}\right) - A\left(e^{-i\lambda}\right)\right|^2 f_X(\lambda)\, d\lambda \\
&\le M \lim_{n\to\infty} \int \left|A_n\left(e^{-i\lambda}\right) - A\left(e^{-i\lambda}\right)\right|^2 d\lambda \\
&= \lim_{n\to\infty} \|A_n - A\|^2_{L^2[-\pi,\pi]} = 0.
\end{aligned}$$

Corollary 4.3 *Assumptions:*

$$X_t = \sum_{j=0}^{\infty} a_j \varepsilon_{t-j} = \text{Wold decomposition},$$

$$\varepsilon_t = \int_{-\pi}^{\pi} e^{it\lambda} dZ_\varepsilon(\lambda), \sigma_\varepsilon^2 = var(\varepsilon_t),$$

$$A(z) = \sum_{j=0}^{\infty} a_j z^j \ (z \in \mathbb{C}).$$

Then

$$X_t = \int_{-\pi}^{\pi} e^{it\lambda} dZ_X(\lambda)$$

with

$$dZ_X(\lambda) = A\left(e^{-i\lambda}\right) dZ_\varepsilon(\lambda)$$

and

$$\exists f_X(\lambda) = F_X'(\lambda) = \frac{\sigma_\varepsilon^2}{2\pi}\left|A\left(e^{-i\lambda}\right)\right|^2 = \frac{\sigma_\varepsilon^2}{2\pi}\left|\sum_{j=0}^{\infty} a_j e^{-ij\lambda}\right|^2.$$

Proof

$$\exists f_\varepsilon(\lambda) = \frac{\sigma_\varepsilon^2}{2\pi} = M < \infty \ (\lambda \in [-\pi, \pi])$$

and

$$\sum a_j^2 < \infty$$

imply

$$A_n\left(e^{-i\lambda}\right) \underset{L^2(F_\varepsilon)}{\rightarrow} A\left(e^{-i\lambda}\right)$$

and hence

$$X_{t,n} = A_n(B)\,\varepsilon_t \underset{L^2(\Omega)}{\rightarrow} X_t = A(B)\,\varepsilon_t$$

and

$$f_X(\lambda) = \left|A\left(e^{-i\lambda}\right)\right|^2 f_\varepsilon(\lambda).$$

4.5.2 Elimination of Frequency Bands

Problem 4.6

$$X_t = \int_{-\pi}^{\pi} e^{it\lambda} dZ_X(\lambda) \ (t \in \mathbb{Z}),\ \Lambda \subset [-\pi, \pi]$$

$$\Downarrow ?$$

$\exists$linear filter A s.t.:

$$Y_t = A(B) X_t = \int e^{it\lambda} A\left(e^{-i\lambda}\right) dZ_X(\lambda) = \int e^{it\lambda} dZ_Y(\lambda),$$

$$dZ_Y(\lambda) \underset{L^2(F_X)}{=} 1\{\lambda \notin \Lambda\} dZ_X(\lambda).$$

Solution 4.6 Approximate desired function by Fourier series.

Example 4.4 Seasonal differencing:

$$\Lambda = \left\{\pm\frac{2\pi}{12}j, j = 1, 2, \ldots, 6\right\},$$

$$A\left(e^{-i\lambda}\right) = 1 - e^{-i12\lambda}.$$

Then

$$A\left(e^{-i\lambda}\right) = 0\ (\lambda \in \Lambda),$$

however

$$A\left(e^{-i\lambda}\right)\ (\lambda \notin \Lambda) \text{ not constant},$$

so that

$$dZ_Y(\lambda) \neq 1\{\lambda \notin \Lambda\} dZ_X(\lambda).$$

Example 4.5 Low pass filter:

$$\Lambda = [-\pi, -\lambda_0) \cup (\lambda_0, \pi] \ (0 < \lambda_0 < \pi),$$

$$X_t \text{ s.t. } \forall \lambda \in [-\pi, \pi] : \exists f_X(\lambda) = F_X'(\lambda) \leq M < \infty,$$

$$A\left(e^{-i\lambda}\right) = 1\{\lambda \in [-\lambda_0, \lambda_0]\} \in L^2(F_X).$$

Then

$$A\left(e^{-i\lambda}\right) \underset{L^2(F_X)}{=} \sum_{j=-\infty}^{\infty} a_j e^{-i\lambda t}$$

with

$$a_j = \frac{1}{2\pi} \int_{-\lambda_0}^{\lambda_0} e^{ij\lambda} d\lambda = \frac{1}{2\pi} \int_{-\lambda_0}^{\lambda_0} \cos(j\lambda)\, d\lambda,$$

and

$$Y_t = A(B) X_t = \sum_{j=-\infty}^{\infty} a_j X_{t-j} = \int_{-\pi}^{\pi} e^{it\lambda} dZ_Y(\lambda),$$

with

$$dZ_Y(\lambda) = A\left(e^{-i\lambda}\right) dZ_X(\lambda) \underset{L^2(\Omega)}{=} 1\{\lambda \notin \Lambda\}\, dZ_X(\lambda).$$

$$\Rightarrow A(B) = \sum_{j=-\infty}^{\infty} a_j B^j = \text{perfect low pass filter.}$$

However, $a_j \neq 0$ for infinitely many $j \in \mathbb{Z}$.

Conclusion 9

$$A\left(e^{-i\lambda}\right) = \sum_{j=-N}^{N} a_j e^{-ij\lambda} = \textit{infinitely differentiable periodic function.}$$

Therefore, in general, a finite linear filter with desired properties may not exist.

Chapter 5
Spectral Representation of Real Valued Vector Time Series

5.1 Cross-Spectrum and Spectral Representation

Definition 5.1

$$X_t = (X_{t,1}, \ldots, X_{t,m})^T \in \mathbb{R}^m \ (t \in \mathbb{Z})$$

is called weakly stationary or second order stationary, if

$$\forall t \in \mathbb{Z} : E(X_t) = \mu \in \mathbb{R}^m,$$

and

$$\exists \gamma : \mathbb{Z} \to M(m, m, \mathbb{R}), \ k \to \gamma(k) = \left[\gamma_{jl}(k)\right]_{j,l=1,\ldots,m} \text{ s.t.:}$$

$$\forall t, k \in \mathbb{Z} : cov\left(X_{t+k,j}, X_{t,l}\right) = \gamma_{jl}(k) \ \ (j, l = 1, 2, \ldots, m).$$

The functions

$$\gamma(k) = \left[\gamma_{jl}(k)\right]_{j,l=1,\ldots,m} \ (k \in \mathbb{Z})$$

and

$$\rho(k) = \left[\rho_{jl}(k)\right]_{j,l=1,\ldots,m} = \left[\frac{\gamma_{jl}(k)}{\sqrt{\gamma_{jj}(0)\,\gamma_{ll}(0)}}\right]_{j,l=1,\ldots,m} \ (k \in \mathbb{Z})$$

are called autocovariance and autocorrelation function of X_t respectively. Also, for $j \neq l$, γ_{jl} and ρ_{jl} are called cross-autocovariance and cross-autocorrelation function respectively.

J. Beran, *Mathematical Foundations of Time Series Analysis*,
https://doi.org/10.1007/978-3-319-74380-6_5

Remark 5.1 $\gamma(0)$ is positive semidefinite, whereas for $k \neq 0$, $\gamma(k)$ is not necessarily positive semidefinite.

Remark 5.2 For $j \neq l$, it matters whether we define $\gamma_{jl}(k)$ as $cov(X_{t+k,j}, X_{t,l})$ or $cov(X_{t,j}, X_{t+k,l})$.

Lemma 5.1

$$\gamma_{jl}(k) = \gamma_{lj}(-k)$$

Proof The result follows from

$$E\left(X_{t+k,j}X_{t,l}\right) = E\left(X_{t,j}X_{t-k,l}\right) = E\left(X_{t-k,l}X_{t,j}\right).$$

Remark 5.3 In general

$$\gamma_{jl}(k) \neq \gamma_{lj}(k)$$

and

$$\gamma_{jl}(k) \neq \gamma_{jl}(-k).$$

Remark 5.4 For $j \neq l$, it is possible to have

$$\rho_{jl}(0) = 0.$$

Example 5.1

$$\varepsilon_t \in \mathbb{R}\ (t \in \mathbb{Z}),\ E(\varepsilon_t) = 0,\ 0 < cov(\varepsilon_{t+k}, \varepsilon_t) = \sigma_\varepsilon^2 \delta_{0k} < \infty$$

$$X_{t,1} = \varepsilon_t,\ X_{t,2} = X_{t-1,1}$$

$$\Rightarrow \gamma_{11}(0) = \gamma_{22}(0) = \sigma_\varepsilon^2 > 0,\ \ \rho_{12}(0) = 0$$

Example 5.2

$$\varepsilon_t \in \mathbb{R}\ (t \in \mathbb{Z}),\ E(\varepsilon_t) = 0,\ 0 < cov(\varepsilon_{t+k}, \varepsilon_t) = \sigma_\varepsilon^2 \delta_{0k} < \infty$$

$$X_{t,1} = \varepsilon_t,\ X_{t,2} = X_{t,1} + X_{t-1,1}.$$

Then

$$\gamma_{11}(0) = \sigma_\varepsilon^2,\ \gamma_{22}(0) = 2\sigma_\varepsilon^2,$$

$$\gamma_{12}(1) = E\left[\varepsilon_{t+1}\left(\varepsilon_t + \varepsilon_{t-1}\right)\right] = 0,$$

$$\gamma_{21}(1) = E\left[\left(\varepsilon_{t+1} + \varepsilon_t\right)\varepsilon_t\right] = \sigma_\varepsilon^2.$$

Definition 5.2 Let

$$F:\mathbb{R}\to M\left(m,m,\mathbb{C}\right).$$

Then

$$F=\text{nondecreasing function}$$

$$\Leftrightarrow$$

$$\forall\lambda_2\geq\lambda_1:F\left(\lambda_2\right)-F\left(\lambda_1\right)\geq 0\text{ (i.e. positive semidefinite)}$$

Definition 5.3 Let

$$F:\mathbb{R}\to M\left(m,m,\mathbb{C}\right)\text{ s.t.:}$$

$$F\text{ right-continuous, nondecreasing,}$$

$$F\left(-\infty\right)=0,\ \max_{1\leq j,l\leq m}\left|F_{jl}\left(\infty\right)\right|<\infty.$$

Then F is called a generalized cumulative distribution function (gcdf).

Theorem 5.1 *Assumption:*

$$\gamma:\mathbb{Z}\to M\left(m,m,\mathbb{R}\right).$$

Then

$$\exists X_t\in\mathbb{R}^m\ (t\in\mathbb{Z})\text{ weakly stationary with acf }\gamma$$

$$\Leftrightarrow$$

$$\exists F:\mathbb{R}\to M\left(m,m,\mathbb{C}\right)\text{ gcdf s.t.:}$$

$$F\left(-\pi\right)=0,\ F\left(\infty\right)=F\left(\pi\right),$$

$$\forall k\in\mathbb{Z}:\gamma\left(k\right)=\int_{-\pi}^{\pi}e^{ik\lambda}dF\left(\lambda\right)$$

Proof Analogous to the univariate case ($m=1$); see e.g. Brockwell and Davis (1991), Hannan (1970) and Gihman and Skorohod (1974).

Definition 5.4 Let $X_t\in\mathbb{R}^m$ ($t\in\mathbb{Z}$) be weakly stationary with F_X as in Theorem 5.1. Then $F_X=\left(F_{jl}\right)_{j,l=1,\ldots,m}$ is called spectral distribution (sdf) of X_t. Also, for $j\neq l$,

F_{jl} is called cross-spectrum between $X_{t,j}$ and $X_{t,l}$. If $\exists f_X = F'_X$, then

$$f_X(\lambda) = \begin{pmatrix} f_{11}(\lambda) & \cdots & f_{1m}(\lambda) \\ \vdots & \ddots & \vdots \\ f_{m1}(\lambda) & \cdots & f_{mm}(\lambda) \end{pmatrix} \in M(m, m, \mathbb{C})$$

is called spectral density function of X_t. Also, for $j \neq l$, f_{jl} is called cross-spectral density function between $X_{t,j}$ and $X_{t,l}$.

Definition 5.5 Let

$$Z(\lambda) \in \mathbb{C}^m \ (\lambda \in [-\pi, \pi]) \text{ stochastic process s.t.:}$$

$$E[Z(\lambda)] = 0, E\left[\|Z(\lambda)\|^2\right] < \infty \ (\lambda \in [-\pi, \pi]),$$

$$\lim_{\delta \downarrow 0} E\left\{\|Z(\lambda + \delta) - Z(\lambda)\|^2\right\} = 0 \ (\lambda \in [-\pi, \pi]).$$

Then Z is called a right-continuous process. Also, if, for all $\lambda_1 < \lambda_2 \leq \lambda_3 < \lambda_3$,

$$E\left[(Z(\lambda_2) - Z(\lambda_1))\left(\overline{Z(\lambda_4) - Z(\lambda_3)}\right)^T\right] = 0 \in M(m, m, \mathbb{C}),$$

then Z is called an orthogonal increment process.

Theorem 5.2 *Assumptions:*

$$X_t \in \mathbb{R}^m \ (t \in \mathbb{Z}) \textit{ weakly stationary with sdf } F_X.$$

Then

$$\exists! Z_X(\lambda) \in \mathbb{C}^m \ (\lambda \in [-\pi, \pi]) \textit{ right-continuous orthogonal increment process s.t.:}$$

$$Z_X(-\pi) = 0,$$

for $\lambda_2 \geq \lambda_1$,

$$F_X(\lambda_2) - F_X(\lambda_1) = E\left[(Z_X(\lambda_2) - Z_X(\lambda_1))\overline{(Z_X(\lambda_2) - Z_X(\lambda_1))}^T\right],$$

$$\forall t \in \mathbb{Z}: X_t \underset{a.s.}{=} \int_{-\pi}^{\pi} e^{it\lambda} dZ_X(\lambda).$$

Proof Analogous to the univariate case ($m = 1$), see e.g. Brockwell and Davis (1991).

Remark 5.5

$$\begin{aligned}\gamma_{jl}(k) &= E\left(X_{t+k,j}X_{t,l}\right)\\ &= \int_{-\pi}^{\pi}\int_{-\pi}^{\pi} e^{i(t+k)\lambda}e^{-it\nu}E\left[dZ_j(\lambda)\overline{dZ_l(\nu)}\right]\\ &= \int_{-\pi}^{\pi} e^{ik\lambda}E\left[dZ_j(\lambda)\overline{dZ_l(\lambda)}\right].\end{aligned}$$

so that

$$\begin{aligned}dF_{jl}(\lambda) &= E\left[dZ_j(\lambda)\overline{dZ_l(\lambda)}\right]\\ &= cov\left(dZ_j(\lambda), dZ_l(\lambda)\right)\end{aligned}$$

and

$$\underbrace{dF(\lambda)}_{m\times m} = E\left[dZ(\lambda)\overline{dZ^T(\lambda)}\right] = var\left(dZ(\lambda)\right).$$

Definition 5.6 Given a function

$$\gamma : \mathbb{Z} \to M(m, m, \mathbb{R}),$$

its discrete Fourier transform $\mathcal{F}_\gamma$ is defined by

$$\mathcal{F}_\gamma : [-\pi, \pi] \to M(m, m, \mathbb{C}), \lambda \to \mathcal{F}_\gamma(\lambda) = \frac{1}{2\pi}\sum_{k=-\infty}^{\infty} e^{-ik\lambda}\gamma(k).$$

Lemma 5.2 *Assumptions:*

$X_t \in \mathbb{R}^m$ *($t \in \mathbb{Z}$) weakly stationary with sdf* F_X,

$$\forall j \in \{1, \ldots, m\} : \sum_{k=-\infty}^{\infty} \left|\gamma_{jj}(k)\right| < \infty.$$

Then

$$\exists f_X = F_X'$$

and

$$f_X = \mathcal{F}_\gamma(\lambda) \ (\lambda \in [-\pi, \pi]).$$

Proof Analogous to the univariate case ($m = 1$), for a detailed proof see e.g. Brockwell and Davis (1991).

Lemma 5.3 *Let*

$$f_X = \mathcal{F}_\gamma.$$

Then

$$f(\lambda) = \overline{f^T(\lambda)}, \text{ i.e. } f = \text{Hermitian matrix.}$$

Proof

$$f_{jl}(\lambda) = \frac{1}{2\pi} \sum_{k=-\infty}^{\infty} e^{-ik\lambda} \gamma_{jl}(k),$$

$$\begin{aligned} f_{lj}(\lambda) &= \frac{1}{2\pi} \sum_{k=-\infty}^{\infty} e^{-ik\lambda} \gamma_{lj}(k) = \frac{1}{2\pi} \sum_{k=-\infty}^{\infty} e^{-ik\lambda} \gamma_{jl}(-k) \\ &= \frac{1}{2\pi} \sum_{k=-\infty}^{\infty} e^{ik\lambda} \gamma_{jl}(k) = \overline{f_{jl}(\lambda)}. \end{aligned}$$

Lemma 5.4 *Let*

$$f_X = (f_{jl})_{j,l=1,\ldots,m} = \mathcal{F}_\gamma.$$

Then

$$f_{jl}(\lambda) = f_{jl}(-\lambda)$$

$$\Leftrightarrow$$

$$\operatorname{Im} f_{jl}(\lambda) = 0$$

$$\Leftrightarrow$$

$$\gamma_{jl}(-k) = \gamma_{jl}(k) \ (k \in \mathbb{Z})$$

Proof

$$f_{jl}(\lambda) = \frac{1}{2\pi} \sum_{k=-\infty}^{\infty} e^{-ik\lambda} \gamma_{jl}(k) = \mathrm{Re} f_{jl}(\lambda) + i \,\mathrm{Im} f_{jl}(\lambda)$$

$$\mathrm{Re} f_{jl}(\lambda) = \frac{1}{2\pi} \sum_{k=-\infty}^{\infty} \gamma_{jl}(k) \cos k\lambda, \ \mathrm{Im} f_{jl}(\lambda) = -\frac{1}{2\pi} \sum_{k=-\infty}^{\infty} \gamma_{jl}(k) \sin k\lambda$$

$$\begin{aligned} f_{jl}(-\lambda) &= \mathrm{Re} f_{jl}(-\lambda) + i \,\mathrm{Im} f_{jl}(-\lambda) \\ &= \mathrm{Re} f_{jl}(\lambda) - i \,\mathrm{Im} f_{jl}(\lambda) \end{aligned}$$

Hence

$$f_{jl}(-\lambda) = f_{jl}(\lambda) \Leftrightarrow \mathrm{Im} f_{jl}(\lambda) = 0$$

Definition 5.7

$$\begin{aligned} c_{jl}(\lambda) &= \mathrm{Re} f_{jl}(\lambda) \\ &= \frac{1}{2\pi} \sum_{k=-\infty}^{\infty} \frac{\gamma_{jl}(k) + \gamma_{jl}(-k)}{2} \cos k\lambda \\ &= \text{cospectrum} \\ &= \text{coincident spectral density} \end{aligned}$$

$$\begin{aligned} q_{jl}(\lambda) &= -\mathrm{Im} f_{jl}(\lambda) \\ &= -\frac{1}{2\pi} \sum_{k=-\infty}^{\infty} \frac{\gamma_{jl}(k) - \gamma_{jl}(-k)}{2} \sin k\lambda \\ &= \text{quadrature spectrum} \\ &= \text{quadrature spectral density} \end{aligned}$$

Lemma 5.5

$$c_{jl}(-\lambda) = c_{jl}(\lambda), \, c_{jl}(\lambda) = c_{lj}(\lambda),$$

$$q_{jl}(-\lambda) = -q_{jl}(\lambda), \, q_{jl}(\lambda) = -q_{lj}(\lambda).$$

Proof Follows directly from the definitions.

Lemma 5.6 *Assumptions:*

$$X_t \in \mathbb{R}^m \ (t \in \mathbb{Z}) \text{ weakly stationary, } \gamma_X = (\gamma_{jl})_{j,l=1,\ldots,m}$$

$$\sum_{k=-\infty}^{\infty} |\gamma_{jj}(k)| < \infty \ (j = 1, \ldots, m)$$

Then

$$\exists f = \mathscr{F}_\gamma(\lambda)$$

and

$$\forall \lambda \in [-\pi, \pi] : f(\lambda) \geq 0, \text{ i.e. } f(\lambda) \text{ positive semidefinite.}$$

Proof Since

$$\sum_{k=-\infty}^{\infty} |\gamma_{jj}(k)| < \infty,$$

Lemma 5.2 implies

$$f_X(\lambda) = \mathscr{F}_\gamma(\lambda) = \frac{1}{2\pi} \sum_{k=-\infty}^{\infty} e^{-ik\lambda} \gamma_X(k).$$

Let

$$a \in \mathbb{C}^m, \ Y_t = \sum_{j=1}^{m} a_j X_{t,j}.$$

Then

$$Y_t \in \mathbb{C} \ (t \in \mathbb{Z}) \text{ weakly stationary,}$$

$$\gamma_Y(k) = \sum_{j,l=1}^{m} a_j \overline{a_l} \gamma_{jl}(k).$$

In particular,

$$
\begin{aligned}
0 &\le \frac{1}{2\pi n}\sum_{s,t=1}^{n}\gamma_Y(t-s)\,e^{-it\lambda}e^{is\lambda} = \frac{1}{2\pi}\sum_{k=-(n-1)}^{(n-1)}\left(1-\frac{|k|}{n}\right)\gamma_Y(k)\,e^{-ik\lambda}\\
&= \frac{1}{2\pi}\sum_{k=-(n-1)}^{(n-1)}\left(1-\frac{|k|}{n}\right)\left\{\sum_{j,l=1}^{m}a_j\overline{a_l}\gamma_{jl}(k)\right\}e^{-ik\lambda}\\
&= \sum_{j,l=1}^{m}a_j\overline{a_l}\left\{\frac{1}{2\pi}\sum_{k=-(n-1)}^{(n-1)}\left(1-\frac{|k|}{n}\right)\gamma_{jl}(k)\,e^{-ik\lambda}\right\}.
\end{aligned}
$$

Now

$$
\sum_{k=-\infty}^{\infty}\left|\gamma_{jj}(k)\right| < \infty\ (j=1,\ \ldots,p) \Rightarrow \left|k\gamma_{jl}(k)\right| \text{ Cesaro summable with limit 0}
$$

$$
\Rightarrow \lim_{n\to\infty}\frac{1}{2\pi}\sum_{k=-(n-1)}^{(n-1)}\left(1-\frac{|k|}{n}\right)\gamma_{jl}(k)\,e^{-ik\lambda} = f_{jl}(\lambda).
$$

Thus

$$
\begin{aligned}
0 &\le \lim_{n\to\infty}\sum_{j,l=1}^{m}a_j\overline{a_l}\left\{\frac{1}{2\pi}\sum_{k=-(n-1)}^{(n-1)}\left(1-\frac{|k|}{n}\right)\gamma_{jl}(k)\,e^{-ik\lambda}\right\}\\
&= \sum_{j,l=1}^{m}a_j\overline{a_l}f_{jl}(\lambda).
\end{aligned}
$$

Lemma 5.7 *Assumption:*

$$
\sum_{k=-\infty}^{\infty}\gamma_{jj}(k)\Big| < \infty\ (j=1,\ldots,m)
$$

Then

$$
\forall j,l=1,\ldots,m,\ \forall\lambda\in[-\pi,\pi]:
$$

$$
\left|f_{jl}(\lambda)\right|^2 \le f_{jj}(\lambda)\,f_{ll}(\lambda)
$$

Proof

$$f(\lambda) \geq 0$$

implies

$$0 \leq g(\lambda) = \begin{pmatrix} f_{jj}(\lambda) & f_{jl}(\lambda) \\ f_{lj}(\lambda) & f_{ll}(\lambda) \end{pmatrix}$$

with

$$f_{lj}(\lambda) = \overline{f_{jl}(\lambda)},$$

and hence

$$0 \leq \det g(\lambda) = f_{jj}(\lambda) f_{ll}(\lambda) - f_{jl}(\lambda) \overline{f_{jl}(\lambda)}.$$

Remark 5.6 The assumption

$$\sum_{k=-\infty}^{\infty} \left| \gamma_{jj}(k) \right| < \infty \tag{5.1}$$

is sufficient but not necessary to obtain the results in this section.

5.2 Coherence and Phase

Definition 5.8

$$A_{jl}(\lambda) = \left| f_{jl}(\lambda) \right| = \text{cross-amplitude spectrum}$$

$$\varphi_{jl}(\lambda) = \arctan \frac{\operatorname{Im} f_{jl}(\lambda)}{\operatorname{Re} f_{jl}(\lambda)} = \text{phase spectrum}$$

Remark 5.7

$$f_{jl}(\lambda) = A_{jl}(\lambda) \exp\left(i\varphi_{jl}(\lambda)\right) \ (\varphi_{jl} \in [0, 2\pi),\ A_{jl} \in \mathbb{R}_{+}),$$

$$A_{jl}(\lambda) = A_{lj}(\lambda),$$

$$\varphi_{jl}(\lambda) = -\varphi_{lj}(\lambda).$$

Definition 5.9 With the convention $0/0 = 0$,

$$\kappa_{jl}(\lambda) = \frac{|f_{jl}(\lambda)|^2}{f_{jj}(\lambda) f_{ll}(\lambda)} \quad (\lambda \in [-\pi, \pi])$$

is called coherence or coherency function.

Remark 5.8 κ_{jl} is sometimes also called squared coherency.

Lemma 5.8

$$0 \le \kappa_{jl}(\lambda) \le 1$$

Proof

$$0 \le |f_{jl}(\lambda)|^2 \le f_{jj}(\lambda) f_{ll}(\lambda)$$

and

$$f_{jj}(\lambda) \ge 0, f_{ll}(\lambda) \ge 0$$

imply

$$0 \le \kappa_{jl}(\lambda) = \frac{|f_{jl}(\lambda)|^2}{f_{jj}(\lambda) f_{ll}(\lambda)} \le 1.$$

Notation 13 *For* $\delta > 0$,

$$\Delta F^{(\delta)}(\lambda) = F(\lambda + \delta) - F(\lambda),$$

$$\Delta Z^{(\delta)}(\lambda) = Z(\lambda + \delta) - Z(\lambda).$$

Lemma 5.9 *Let*

$$X, Y \in L^2(\Omega; \mathbb{C}),$$

$$E(X) = E(Y) = 0, 0 < var(X), var(Y) < \infty,$$

$$a_{opt} = \underset{a \in \mathbb{C}}{argmin} \|Y - aX\|^2_{L^2(\Omega)}.$$

Then

$$a_{opt} = \frac{\langle Y, X \rangle_{L^2(\Omega)}}{\|X\|^2_{L^2(\Omega)}}.$$

Proof For

$$a = a_1 + ia_2 \ (a_1, a_2 \in \mathbb{R}),$$

$$g(a) = \|Y - aX\|_{L^2(\Omega)}^2 = E\left[(Y - aX)\overline{(Y - aX)}\right]$$
$$= \|Y\|^2 - (a_1 - ia_2) E\left(Y\overline{X}\right) - (a_1 + ia_2) E\left(X\overline{Y}\right) + \left(a_1^2 + a_2^2\right)\|X\|^2$$

The result follows from solving

$$0 = \frac{\partial}{\partial a_1} g(a), 0 = \frac{\partial}{\partial a_1} g(a).$$

Corollary 5.1 *Assumptions:*

$$X_t \in \mathbb{R}^m \ (t \in \mathbb{Z}), X_t = \int_{-\pi}^{\pi} e^{it\lambda} dZ_X(\lambda)$$

$$Z_X = (Z_1, \ldots, Z_m)^T, F_X = \left(F_{jl}\right)_{j,l=1,2,\ldots,m}$$

$$0 < \lambda < \pi, \delta \in (0, \pi - \lambda]$$

$$a_{jl}^{(\delta)}(\lambda) = \frac{\Delta F_{jl}^{(\delta)}(\lambda)}{\Delta F_{ll}^{(\delta)}(\lambda)}$$

Then

$$\Delta Z_j^{(\delta)}(\lambda) = a_{jl}^{(\delta)}(\lambda)\,\Delta Z_l^{(\delta)}(\lambda) + \Delta E_{jl}^{(\delta)}(\lambda)$$

with

$$cov\left(\Delta Z_l^{(\delta)}(\lambda), \Delta E_{jl}^{(\delta)}(\lambda)\right) = 0,$$

and

$$corr\left(\Delta Z_j^{(\delta)}(\lambda), \Delta Z_l^{(\delta)}(\lambda)\right) = \frac{\Delta F_{jl}^{(\delta)}(\lambda)}{\sqrt{\Delta F_{jj}^{(\delta)}(\lambda)\,\Delta F_{ll}^{(\delta)}(\lambda)}}.$$

Proof

1) Case 1:

$$\min\left\{\Delta F_{jj}^{(\delta)}(\lambda), \Delta F_{ll}^{(\delta)}(\lambda)\right\} = 0$$

$$\Rightarrow \Delta F_{jl}^{(\delta)}(\lambda) = 0$$

In this case, the results follow from the convention $0/0 := 0$:

$$\Delta F_{jl}^{(\delta)}(\lambda) = 0 \Rightarrow a_{jl}^{(\delta)}(\lambda) = 0, corr\left(\Delta Z_j^{(\delta)}(\lambda), \Delta Z_l^{(\delta)}(\lambda)\right) = 0$$

2) Case 2:

$$\min\left\{\Delta F_{jj}^{(\delta)}(\lambda), \Delta F_{ll}^{(\delta)}(\lambda)\right\} > 0$$

Notation:

$$P_{\Delta Z_l^{(\delta)}}\left(\Delta Z_j^{(\delta)}(\lambda)\right) = \text{orthogonal projection of } \Delta Z_j^{(\delta)}(\lambda) \text{ on } \Delta Z_l^{(\delta)}(\lambda).$$

Then Lemma 5.9 implies

$$P_{\Delta Z_l^{(\delta)}}\left(\Delta Z_j^{(\delta)}(\lambda)\right) = a_{\text{opt}} \Delta Z_l^{(\delta)}(\lambda)$$

with

$$a_{\text{opt}} = \frac{\left\langle \Delta Z_j^{(\delta)}(\lambda), \Delta Z_l^{(\delta)}(\lambda)\right\rangle}{\left\|\Delta Z_l^{(\delta)}(\lambda)\right\|^2} = \frac{\Delta F_{jl}^{(\delta)}(\lambda)}{\Delta F_{ll}^{(\delta)}(\lambda)} = a_{jl}^{(\delta)}(\lambda).$$

Hence

$$\Delta Z_j^{(\delta)}(\lambda) = a_{jl}^{(\delta)}(\lambda)\, \Delta Z_l^{(\delta)}(\lambda) + E_{jl}^{(\delta)}(\lambda)$$

with

$$\left\langle \Delta Z_l^{(\delta)}(\lambda), E_{jl}^{(\delta)}(\lambda)\right\rangle = cov\left(\Delta Z_l^{(\delta)}(\lambda), E_{jl}^{(\delta)}(\lambda)\right) = 0,$$

$$\left\langle \Delta Z_j^{(\delta)}(\lambda), \Delta Z_l^{(\delta)}(\lambda)\right\rangle = a_{jl}^{(\delta)}(\lambda) \left\|\Delta Z_l^{(\delta)}(\lambda)\right\|^2 = a_{jl}^{(\delta)}(\lambda)\, \Delta F_{ll}^{(\delta)}(\lambda),$$

and

$$
\begin{aligned}
corr\left(\Delta Z_j^{(\delta)}(\lambda), \Delta Z_l^{(\delta)}(\lambda)\right) &= \frac{\left\langle \Delta Z_j^{(\delta)}(\lambda), \Delta Z_l^{(\delta)}(\lambda)\right\rangle}{\left\| \Delta Z_j^{(\delta)}(\lambda)\right\| \left\| \Delta Z_l^{(\delta)}(\lambda)\right\|} \\
&= a_{jl}^{(\delta)}(\lambda) \frac{\Delta F_{ll}^{(\delta)}(\lambda)}{\sqrt{\Delta F_{jj}^{(\delta)}(\lambda)\, \Delta F_{ll}^{(\delta)}(\lambda)}} \\
&= \frac{\Delta F_{jl}^{(\delta)}(\lambda)}{\Delta F_{ll}^{(\delta)}(\lambda)} \frac{\Delta F_{ll}^{(\delta)}(\lambda)}{\sqrt{\Delta F_{jj}^{(\delta)}(\lambda)\, \Delta F_{ll}^{(\delta)}(\lambda)}} \\
&= \frac{\Delta F_{jl}^{(\delta)}(\lambda)}{\sqrt{\Delta F_{jj}^{(\delta)}(\lambda)\, \Delta F_{ll}^{(\delta)}(\lambda)}}.
\end{aligned}
$$

Corollary 5.2 *Assumptions as in Corollary 5.1. Furthermore,*

$$\exists f_X = \left(f_{jl}\right)_{j,l=1,\ldots,m} = F_X' \in M(m, m, \mathbb{C}),$$

$$a_{jl}(\lambda) = \frac{f_{jl}(\lambda)}{f_{ll}(\lambda)} \; (\lambda \in [-\pi, \pi]\,).$$

Then

$$a_{jl}(\lambda) = \lim_{\delta \downarrow 0} a_{jl}^{(\delta)}(\lambda),$$

$$\lim_{\delta \downarrow 0} corr\left(\Delta Z_j^{(\delta)}(\lambda), \Delta Z_l^{(\delta)}(\lambda)\right) = \frac{f_{jl}(\lambda)}{\sqrt{f_{jj}(\lambda) f_{ll}(\lambda)}} = a_{jl}(\lambda) \sqrt{\frac{f_{ll}(\lambda)}{f_{jj}(\lambda)}},$$

$$\lim_{\delta \downarrow 0} corr\left(\Delta Z_l^{(\delta)}(\lambda), \Delta E_{jl}^{(\delta)}(\lambda)\right) = 0.$$

Proof

a)

$$\lim_{\delta \downarrow 0} a_{jl}^{(\delta)}(\lambda) = \lim_{\delta \downarrow 0} \frac{\left[F_{jl}(\lambda + \delta) - F_{jl}(\lambda)\right]/\delta}{\left[F_{ll}(\lambda + \delta) - F_{ll}(\lambda)\right]/\delta} = \frac{f_{jl}(\lambda)}{f_{ll}(\lambda)}$$

b)

$$cov\left(\Delta Z_l^{(\delta)}(\lambda), E_{jl}^{(\delta)}(\lambda)\right) = \left\langle \Delta Z_l^{(\delta)}(\lambda), E_{jl}^{(\delta)}(\lambda)\right\rangle = 0$$

$$\Rightarrow \lim_{\delta\downarrow 0} corr\left(\Delta Z_l^{(\delta)}(\lambda), E_{jl}^{(\delta)}(\lambda)\right) = 0.$$

c)

$$\lim_{\delta\downarrow 0} \frac{\left\langle \Delta Z_j^{(\delta)}(\lambda), \Delta Z_l^{(\delta)}(\lambda)\right\rangle}{\delta} = \lim_{\delta\downarrow 0} \frac{F_{jl}(\lambda+\delta) - F_{jl}(\lambda)}{\delta} = f_{jl}(\lambda),$$

$$\lim_{\delta\downarrow 0} \frac{\left\| \Delta Z_{jj}^{(\delta)}(\lambda)\right\|^2}{\delta} = \lim_{\delta\downarrow 0} \frac{F_{jj}(\lambda+\delta) - F_{jj}(\lambda)}{\delta} = f_{jj}(\lambda),$$

$$\lim_{\delta\downarrow 0} \frac{\left\| \Delta Z_{ll}^{(\delta)}(\lambda)\right|^2}{\delta} = \lim_{\delta\downarrow 0} \frac{F_{ll}(\lambda+\delta) - F_{ll}(\lambda)}{\delta} = f_{ll}(\lambda).$$

This implies

$$\lim_{\delta\downarrow 0} corr\left(\Delta Z_j^{(\delta)}(\lambda), \Delta Z_l^{(\delta)}(\lambda)\right) = \lim_{\delta\downarrow 0} \frac{\Delta F_{jl}^{(\delta)}(\lambda)/\delta}{\sqrt{\Delta F_{jj}^{(\delta)}(\lambda)/\delta}\sqrt{\Delta F_{ll}^{(\delta)}(\lambda)/\delta}}$$

$$= \frac{f_{jl}(\lambda)}{\sqrt{f_{jj}(\lambda) f_{ll}(\lambda)}}.$$

Remark 5.9 In the infinitesimal notation we write

$$corr\left(dZ_j(\lambda), dZ_l(\lambda)\right) = \frac{f_{jl}(\lambda)}{\sqrt{f_{jj}(\lambda) f_{ll}(\lambda)}} = a_{jl}(\lambda)\sqrt{\frac{f_{ll}(\lambda)}{f_{jj}(\lambda)}},$$

Corollary 5.3 *Under the assumptions of Corollary 5.2,*

$$\kappa_{jl}(\lambda) = \left|corr\left(dZ_j(\lambda), dZ_l(\lambda)\right)\right|^2,$$

$$\kappa_{jl}(\lambda) = 1 \Leftrightarrow dZ_j(\lambda) = a_{jl}(\lambda)\, dZ_l(\lambda),$$

$$\kappa_{jl}(\lambda) = 0 \Leftrightarrow corr\left(dZ_j(\lambda), dZ_l(\lambda)\right) = 0.$$

Remark 5.10 By definition of Z,

$$corr\left(dZ_j(\lambda), dZ_l(\nu)\right) = 0\ (\lambda \neq \nu, j, l = 1, \ldots, m).$$

Thus, linear dependence between $X_{t,j}$ and $X_{t,l}$ is only possible at identical frequencies λ.

Notation 14

$$\tau_{jl}(\lambda) = \frac{\varphi_{jl}(\lambda)}{\lambda},\ t_{jl}(\lambda) = t + \tau_{jl}(\lambda).$$

Corollary 5.4 *Under the assumptions of Corollary 5.2, we have, for* $j, l = 1, \ldots, m$,

$$X_{t,j} = X_{t,jl} + e_{t,jl}$$

with

$$X_{t,jl} = \int_{-\pi}^{\pi} \exp\left(it_{jl}(\lambda)\lambda\right) \cdot \left|a_{jl}(\lambda)\right| dZ_l(\lambda),$$

$$e_{t,jl} = \int_{-\pi}^{\pi} \exp(it\lambda) \cdot dE_{jl}(\lambda),$$

and

$$cov\left(dZ_l(\lambda), dE_{jl}(\lambda)\right) = 0.$$

Remark 5.11 $X_{t,jl}$ is obtained from $X_{t,l}$ as follows: (1) decompose $X_{t,l}$ into periodic contributions $\exp(it\lambda)dZ_l(\lambda)$ $(\lambda \in [-\pi, \pi])$; (2) for each λ, multiply the random amplitude $dZ_l(\lambda)$ by the factor $\left|a_{jl}(\lambda)\right| = \left|f_{jl}(\lambda)\right|/f_{jj}(\lambda) \in \mathbb{R}$; (3) for each λ, shift time t by $\tau_{jl}(\lambda) = \varphi_{jl}(\lambda)\lambda^{-1}$ to obtain the new frequency dependent time

$$t_{jl}(\lambda) = t + \tau_{jl}(\lambda).$$

The interpretation of φ_{jl} and t_{jl} is easiest, when t_{jl} is piecewise constant:

Corollary 5.5 *Assumptions as in Corollary 5.4, and*

$$0 < a < b \leq \pi,\ \alpha \in \mathbb{R},$$

$$t_{jl}(\lambda) = t + \alpha\ (\lambda \in [a, b]).$$

Then

$$\exists \varphi'_{jl}(\lambda)\ (\lambda \in (a, b))$$

and

$$t_{jl}(\lambda) = t + \varphi'_{jl}(\lambda) \ \ (\lambda \in (a, b)).$$

Proof

$$t_{jl}(\lambda) = t + \varphi_{jl}(\lambda)\lambda^{-1} = t + \alpha \ (\lambda \in [a, b])$$

implies

$$\varphi_{jl}(\lambda) = \alpha\lambda \ (\lambda \in [a, b])$$

and hence

$$\exists \varphi'_{jl}(\lambda) = \alpha \ (\lambda \in (a, b))$$

and

$$t_{jl}(\lambda) = t + \alpha = t + \varphi'_{jl}(\lambda) \ \ (\lambda \in (a, b)).$$

Remark 5.12 Corollary 5.5 means that for piecewise constant time shifts t_{jl},

$$\varphi'_{jl}(\lambda) = \text{local time shift at frequency } \lambda.$$

More generally, if t_{jl} is a continuous function of λ, then we may approximate t_{jl} uniformly by piecewise constant functions. In this sense, $\varphi'_{jl}(\lambda)$ may be interpreted as a local time shift of the component dZ_l at frequency λ.

Definition 5.10 $\varphi'_{jl}(\lambda)$ is called phase shift or group delay at frequency λ.

Remark 5.13

$$\varphi'_{jl}(\lambda) = \text{time shift of } dZ_j(\lambda) \ \text{ w.r.t. } dZ_l(\lambda)$$

$$\varphi'_{jl}(\lambda) > 0 \Rightarrow dZ_j(\lambda) \ \text{ ahead of } dZ_l(\lambda),$$

$$\varphi'_{jl}(\lambda) = 0 \Rightarrow dZ_j(\lambda), dZ_l(\lambda) \ \text{ synchronous},$$

$$\varphi'_{jl}(\lambda) < 0 \Rightarrow dZ_j(\lambda) \ \text{ behind } dZ_l(\lambda).$$

Remark 5.14 The interpretation of $\varphi'_{jl}(\lambda)$ as a time shift is clear, if time t is continuous. On the other hand, strictly speaking, for discrete time $t \in \mathbb{Z}$ the interpretation of $\varphi'_{jl}(\lambda)$ as a time shift is straightforward only, if $\varphi'_{jl}(\lambda)$ is an integer or at least close to an integer.

Remark 5.15 κ_{jl} only measures linear dependence. Thus,

$$\kappa_{jl} \equiv 0 \nRightarrow dZ_j, dZ_l \text{ independent}$$

Remark 5.16

$$\kappa_{jl}(\lambda) \equiv 1 \nRightarrow X_{t,j} = X_{t,l}$$

$$\gamma_{jl}(0) = 0 \nRightarrow \kappa_{jl} \equiv 0$$

Coherence is a finer measure of dependence than correlation, because time shifts can be applied without changing κ_{jl}.

Example 5.3

$$\varepsilon_t \ (t \in \mathbb{Z}), E(\varepsilon_t) = 0, cov(\varepsilon_s, \varepsilon_t) = \delta_{st}\sigma_\varepsilon^2 < \infty$$

$$X_{t,1} = \varepsilon_{t+1}, X_{t,2} = \varepsilon_t$$

Then

$$\gamma_{12}(k) = \sigma_\varepsilon^2 \delta_{-1k},$$

$$f_{11}(\lambda) = f_{22}(\lambda) = \frac{\sigma_\varepsilon^2}{2\pi},$$

$$f_{12}(\lambda) = \frac{1}{2\pi} \sum_{k=-\infty}^{\infty} \gamma_X(k) e^{-ik\lambda} = \frac{\sigma_\varepsilon^2}{2\pi} e^{i\lambda}.$$

Hence

$$\gamma_{12}(0) = cov(X_{t,1}, X_{t,2}) = 0,$$

but

$$\kappa_{12}(\lambda) = \frac{|f_{12}(\lambda)|^2}{f_{11}(\lambda) f_{22}(\lambda)} \equiv 1.$$

Moreover,

$$\varphi_{12}(\lambda) = \arctan \frac{\mathrm{Im} f_{12}(\lambda)}{\mathrm{Re} f_{12}(\lambda)} = \lambda \Rightarrow \varphi'_{12}(\lambda) \equiv 1,$$

$$t_{12}(\lambda) = t + \varphi_{12}(\lambda)\lambda^{-1} = t + 1,$$

which means that for all λ, $dZ_1(\lambda)$ is ahead of $dZ_2(\lambda)$ by the time shift $\varphi'_{12}(\lambda) = 1$.

Example 5.4 Let

$$X_{t,1} \in \mathbb{R}\ (t \in \mathbb{Z}) \text{ weakly stationary}, E\left(X_{t,1}\right) = 0, \exists f_{11} = F'_{11} > 0, a \in \mathbb{R},$$

$$e_t\ (t \in \mathbb{Z}) \text{ weakly stationary}, E\left(e_t\right) = 0, \exists f_e = F'_e,$$

$$a \in \mathbb{R}, a \neq 0,$$

$$\left(e_t\right)_{t\in\mathbb{Z}}, \left(X_{t,1}\right)_{t\in\mathbb{Z}} \text{ uncorrelated},$$

$$X_{t,2} = aX_{t+3,1} + e_t.$$

Then

$$\gamma_{22}(k) = a^2\gamma_{11}(k) + \gamma_e(k),$$

$$\gamma_{12}(k) = cov\left(X_{t+k,1}, aX_{t+3,1} + e_t\right) = a\gamma_{11}(k-3),$$

$$f_{22} = a^2 f_{11} + f_e, f_{12}(\lambda) = e^{-i3\lambda} a f_{11}(\lambda).$$

Hence

$$\varphi_{12}(\lambda) = -3\lambda, \varphi'_{12}(\lambda) = -3 = \text{global time shift},$$

$$\kappa_{12}(\lambda) = \frac{\left|f_{12}(\lambda)\right|^2}{f_{11}(\lambda) f_{22}(\lambda)} = \frac{\left|e^{-i3\lambda} a f_{11}(\lambda)\right|^2}{f_{11}(\lambda)\cdot\left[a^2 f_{11}(\lambda) + f_e(\lambda)\right]} = \frac{1}{1 + a^{-2} f_e(\lambda)/f_{11}(\lambda)}$$

so that

$$\lim_{a\to\infty} \kappa_{12}(\lambda) = \lim_{f_e(\lambda)/f_{11}(\lambda)\to 0} \kappa_{12}(\lambda) = 1$$

and

$$\lim_{a\to 0} \kappa_{12}(\lambda) = \lim_{f_e(\lambda)/f_{11}(\lambda)\to\infty} \kappa_{12}(\lambda) = 0.$$

In contrast

$$\lim_{a\to\infty} corr\left(X_{t,1}, X_{t,2}\right) = \lim_{a\to\infty} \frac{a\gamma_{11}(-3)}{\sqrt{\gamma_{11}(0)\left[a^2\gamma_{11}(0) + \gamma_e(0)\right]}}$$

$$= \lim_{a\to\infty} \frac{\gamma_{11}(-3)}{\sqrt{\gamma_{11}(0)\left[\gamma_{11}(0) + a^{-2}\gamma_e(0)\right]}} = \rho_{11}(-3).$$

Theorem 5.3 *Assumptions:*

$$X_{t,1} \in \mathbb{R}\ (t \in \mathbb{Z}) \text{ weakly stationary}, E\left(X_{t,1}\right) = 0,$$

$$\sum_{k=-\infty}^{\infty} \left|\gamma_{11}(k)\right| < \infty,$$

$$a_j \in \mathbb{R}\ (j \in \mathbb{Z}),\ \sum_{j=-\infty}^{\infty} a_j^2 < \infty,\ A\left(e^{-i\lambda}\right) = \sum_{j=-\infty}^{\infty} a_j e^{-ij\lambda},$$

$$X_{t,2} = A(B)\, X_{t,1} = \sum_{s=0}^{\infty} a_s X_{t-s,1} (t \in \mathbb{Z}),\ X_t = \begin{pmatrix} X_{t,1} \\ X_{t,2} \end{pmatrix}$$

Then X_t $(t \in \mathbb{Z})$ is weakly stationary with spectral density $f = (f_{jl})_{j,l=1,2}$ given by

$$f_{11}(\lambda) = \frac{1}{2\pi} \sum_{k=-\infty}^{\infty} \gamma_{11}(k)\, e^{-ik\lambda},$$

$$f_{22}(\lambda) = \left|A\left(e^{-i\lambda}\right)\right|^2 f_{11}(\lambda)$$

and

$$f_{12}(\lambda) = A\left(e^{i\lambda}\right) f_{11}(\lambda) = e^{i\varphi_{12}(\lambda)} \left|A\left(e^{-i\lambda}\right)\right| f_{11}(\lambda).$$

Moreover,

$$\kappa_{12} \equiv 1$$

Proof

$$\sum_{k=-\infty}^{\infty} \left|\gamma_{11}(k)\right| < \infty \Rightarrow \exists f(\lambda) = \frac{1}{2\pi} \sum_{k=-\infty}^{\infty} \gamma_{11}(k)\, e^{-ik\lambda}$$

Let

$$A_n(z) = \sum_{j=-n}^{n} a_j z^j\ (z \in \mathbb{C}),$$

$$X_{t,2;n} = A_n(B)\, X_{t,1} = \sum_{j=-n}^{n} a_j X_{t-j,1} = \int_{-\pi}^{\pi} e^{it\lambda} A_n\left(e^{-i\lambda}\right) dZ_{X_1}(\lambda).$$

Then

$$A\left(e^{-i\lambda}\right)\in L^{2}\left(F_{X_1}\right),\ \lim_{n\to\infty}\|A-A_n\|_{L^2(F_{X_1})}=0$$

and

$$\lim_{n\to\infty}\|X_{t,2}-X_{t,2;n}\|_{L^2(\Omega)}=\lim_{n\to\infty}\|A-A_n\|_{L^2(F_{X_1})}=0.$$

Hence

$$X_{t,2}=\int e^{it\lambda}A\left(e^{-i\lambda}\right)dZ_{X_1}(\lambda)$$

and

$$\begin{aligned}\gamma_{12}(k)&=cov\left(X_{t+k,1},X_{t,2}\right)\\&=\left\langle\int e^{i(t+k)\lambda}dZ_{X_1}(\lambda),\int e^{it\nu}A\left(e^{-i\nu}\right)dZ_{X_1}(\nu)\right\rangle\\&=\int\int e^{i(t+k)\lambda}e^{-it\nu}\overline{A\left(e^{-i\nu}\right)}E\left[dZ_{X_1}(\lambda)\overline{dZ_{X_1}(\nu)}\right]\\&=\int e^{ik\lambda}\overline{A\left(e^{-i\lambda}\right)}f_{11}(\lambda)\,d\lambda\\&=\int e^{ik\lambda}A\left(e^{i\lambda}\right)f_{11}(\lambda)\,d\lambda.\end{aligned}$$

Since

$$\gamma_{12}(k)=\int e^{ik\lambda}f_{12}(\lambda)\,d\lambda,$$

and

$$\left|A\left(e^{-i\lambda}\right)\right|^{2}=\left|A\left(e^{i\lambda}\right)\right|^{2},$$

we obtain

$$f_{12}(\lambda)=A\left(e^{i\lambda}\right)f_{11}(\lambda)=e^{i\varphi_{12}(\lambda)}\left|A\left(e^{-i\lambda}\right)\right|f_{11}(\lambda)$$

and

$$\kappa_{XY}(\lambda)=\frac{|f_{12}(\lambda)|^{2}}{f_{11}(\lambda)f_{22}(\lambda)}=\frac{\left|A\left(e^{i\lambda}\right)\right|^{2}f_{11}^{2}(\lambda)}{f_{11}(\lambda)\left|A\left(e^{-i\lambda}\right)\right|^{2}f_{11}(\lambda)}\equiv 1.$$

Remark 5.17 Theorem 5.3 means that, whenever $X_{t,2}$ is obtained from $X_{t,1}$ by applying a linear filter, κ_{12} is identically equal to one. In this sense, κ_{12} is a measure of linear dependence between two time series $X_{t,1}$, $X_{t,2}$. A more general result for multivariate filters is obtained in the following.

Theorem 5.4 *Assumptions:*

$$X_t \in \mathbb{R}^m \ (t \in \mathbb{Z}), \text{ weakly stationary, } E(X_t) = 0, \ \exists f_X = F_X'$$

$$a_j = \left[a_j(r,s)\right]_{r,s=1,\ldots,m} \in M(m,m,\mathbb{R}) \ (j \in \mathbb{Z})$$

$$A_n\left(e^{-i\lambda}\right) = \left[A_n\left(e^{-i\lambda}; r,s\right)\right]_{r,s=1,\ldots,m} = \sum_{j=-n}^{n} a_j e^{-ij\lambda}$$

$$A\left(e^{-i\lambda}\right) = \left[A\left(e^{-i\lambda}; r,s\right)\right]_{r,s=1,\ldots,m} = \sum_{j=-\infty}^{\infty} a_j e^{-ij\lambda}$$

$$Y_t(n) = \left[Y_{t,1}(n), \ldots, Y_{t,m}(n)\right]^T = A_n(B) Y_t$$

$$\forall r,s \in \{1,2,\ldots,m\} : A_n\left(e^{-i\lambda}; r,s\right) \underset{L^2(F_{ss})}{\rightarrow} A\left(e^{-i\lambda}; r,s\right)$$

Then

$$\forall j \in \{1,2,\ldots,m\} : Y_{t,j}(n) \underset{L^2(\Omega)}{\rightarrow} Y_{t,j} = \sum_{l=1}^{m} A(B; j,l) X_{t,l} \ (t \in \mathbb{Z}),$$

$$Y_t = (Y_{t,1}, \ldots, Y_{t,m})^T = A(B) X_t \ (t \in \mathbb{Z}) \text{ weakly stationary, } E(Y_t) = 0,$$

$$f_Y(\lambda) = A\left(e^{-i\lambda}\right) f_X(\lambda) \overline{A^T\left(e^{-i\lambda}\right)}.$$

Proof Weak stationarity follow as in the univariate case. For f_Y, consider

$$\begin{aligned}
\gamma_{jl}(k) &= cov\left(Y_{t+k,j}, Y_{t,l}\right) \\
&= E\left\{\sum_{r=1}^{m} A(B; j,r) X_{t+k,r} \sum_{s=1}^{m} A(B; l,s) X_{t,s}\right\} \\
&= \sum_{r,s=1}^{m} E\left\{\left(\sum_{u=-\infty}^{\infty} a_u(j,r) X_{t+k-u,r}\right)\left(\sum_{v=-\infty}^{\infty} a_v(l,s) X_{t-v,s}\right)\right\} \\
&= \sum_{u,v=-\infty}^{\infty} \sum_{r,s=1}^{m} a_u(j,r) a_v(l,s) \gamma_{rs}(k+v-u)
\end{aligned}$$

The spectral representation of γ_{rs} then leads to

$$
\begin{aligned}
\gamma_{jl}(k) &= \sum_{u,v=-\infty}^{\infty} \int_{-\pi}^{\pi} e^{ik\lambda} \sum_{r,s=1}^{m} a_u(j,r)\, a_v(l,s)\, e^{i(v-u)\lambda} f_{rs}(\lambda)\, d\lambda \\
&= \int_{-\pi}^{\pi} e^{ik\lambda} \sum_{r,s=1}^{m} \left(\sum_{u=-\infty}^{\infty} a_u(j,r)\, e^{-iu\lambda} \right) \left(\sum_{v=-\infty}^{\infty} a_v(l,s)\, e^{iv\lambda} \right) f_{rs}(\lambda)\, d\lambda \\
&= \int_{-\pi}^{\pi} e^{ik\lambda} \sum_{r,s=1}^{m} A\left(e^{-i\lambda}; j, r\right) f_{rs}(\lambda)\, \overline{A\left(e^{-i\lambda}; l, s\right)} d\lambda \\
&= \int_{-\pi}^{\pi} e^{ik\lambda} \left[A\left(e^{-i\lambda}\right) f_X(\lambda)\, \overline{A^T\left(e^{-i\lambda}\right)} \right]_{jl} d\lambda .
\end{aligned}
$$

Hence

$$
f_Y(\lambda) = A\left(e^{-i\lambda}\right) f_X(\lambda)\, \overline{A^T\left(e^{-i\lambda}\right)} .
$$

Chapter 6
Univariate ARMA Processes

6.1 Definition

Definition 6.1 Let $p, q \in \mathbb{N}$,

$$\varphi(z) = 1 - \sum_{j=1}^{p} \varphi_j z^j, \ \psi(z) = 1 + \sum_{j=1}^{q} \psi_j z^j \ (\varphi_j, \psi_j \in \mathbb{R}, z \in \mathbb{C}),$$

$$\varepsilon_t \ (t \in \mathbb{Z}) \text{ iid}, E(\varepsilon_t) = 0, \sigma_\varepsilon^2 = var(\varepsilon_t) < \infty.$$

Then X_t is called an autoregressive moving average process of order (p, m), or an $ARMA(p, q)$ process, if

$$\forall t \in \mathbb{Z} : \varphi(B) X_t \underset{a.s.}{=} \psi(B) \varepsilon_t.$$

If $q = 0$, then X_t is also called an autoregressive process of order p, or $AR(p)$ process. If $p = 0$, then X_t is also called a moving average process of order p, or $MA(q)$ process.

6.2 Stationary Solution

Problem 6.1 Existence of a weakly or strictly stationary solution.

Solution 6.1 Conditions on $\varphi(z)$.

J. Beran, *Mathematical Foundations of Time Series Analysis*,
https://doi.org/10.1007/978-3-319-74380-6_6

Theorem 6.1 *Assumptions:*

$$\varepsilon_t \ (t \in \mathbb{Z}) \ iid, \ E(\varepsilon_t) = 0, \ \sigma_\varepsilon^2 = var(\varepsilon_t) < \infty$$

$$\varphi(z) \neq 0 \ for \ z \ with \ |z| = 1$$

Then

$$\exists! X_t \ stationary \ s.t. \ \forall t \in \mathbb{Z} : \varphi(B) X_t \underset{a.s.}{=} \psi(B) \varepsilon_t.$$

Moreover,

$$X_t = \sum_{j=-\infty}^{\infty} a_j \varepsilon_{t-j} \ (a.s. \ and \ L^2(\Omega)),$$

$$\sum |a_j| < \infty,$$

$$\exists r > 1 \ s.t. \ \exists Laurent \ series \ A(z) = \sum_{j=-\infty}^{\infty} a_j z^j = \frac{\psi(z)}{\varphi(z)} \in \mathbb{C} \ (\frac{1}{r} < |z| < r),$$

$$\exists f_X(\lambda) = F_X'(\lambda) = f_X(\lambda) = \frac{\sigma_\varepsilon^2}{2\pi} \left| \frac{\psi(e^{-i\lambda})}{\varphi(e^{-i\lambda})} \right|^2.$$

Proof W.l.o.g. we may assume $q = 0$, which means

$$\varphi(B) X_t = \varepsilon_t.$$

We will show that

$$X_t = \varphi^{-1}(B) \varepsilon_t.$$

a) Define candidate solution $\tilde{X}_t$:

$$\varepsilon_t = \int_{-\pi}^{\pi} e^{i\lambda t} dZ_\varepsilon(\lambda)$$

has the spectral density

$$f_\varepsilon(\lambda) = \frac{\sigma_\varepsilon^2}{2\pi} < \infty.$$

Define

$$A(e^{-i\lambda}) = \frac{1}{\varphi(e^{-i\lambda})}.$$

Then

$$\varphi\left(e^{-i\lambda}\right) = 1 - \sum_{j=1}^{p} \varphi_j e^{-ij\lambda} = \text{trigonometric polynomial},$$

implies

$$\varphi\left(e^{-i\lambda}\right) \neq 0 \ (\lambda \in [-\pi, \pi]) \Rightarrow \exists \delta : \left|\varphi\left(e^{-i\lambda}\right)\right| > \delta \ (\lambda \in [-\pi, \pi])$$

so that

$$\|A\|^2_{L^2(F_\varepsilon)} = \frac{\sigma_\varepsilon^2}{2\pi} \int_{-\pi}^{\pi} \left|A\left(e^{-i\lambda}\right)\right|^2 d\lambda \le \frac{\sigma_\varepsilon^2}{\delta} < \infty$$

and

$$A\left(e^{-i\lambda}\right) \in L^2\left(F_\varepsilon\right).$$

Therefore

$$\tilde{X}_t = \int_{-\pi}^{\pi} e^{it\lambda} A\left(e^{-i\lambda}\right) dZ_\varepsilon(\lambda) \in L^2(\Omega),$$

$$\tilde{X}_t \text{ weakly stationary},$$

$$f_{\tilde{X}}(\lambda) = \frac{\sigma_\varepsilon^2}{2\pi} \frac{1}{\left|\varphi\left(e^{-i\lambda}\right)\right|^2}.$$

b) Summability:

$$|\varphi(z)| > \delta \ (|z| = 1)$$

$$\Rightarrow \exists r > 1 \text{ s.t. } A(z) = \frac{1}{\varphi(z)} = \sum_{j=-\infty}^{\infty} a_j z^j \ \left(\frac{1}{r} < |z| < r\right)$$

where

$$r = \frac{1}{\limsup_{j\to\infty} \sqrt[j]{|a_j|}} > 1.$$

Hence,

$$\sum_{j=-\infty}^{\infty} |a_j| < \infty.$$

c) Is $\tilde{X}_t$ a solution?: Let

$$A_n\left(e^{-i\lambda}\right) = \sum_{j=-n}^{n} a_j e^{-ij\lambda},$$

$$\tilde{X}_{t,n} = A_n(B)\,\varepsilon_t = \int_{-\pi}^{\pi} e^{it\lambda} A_n\left(e^{-i\lambda}\right) dZ_\varepsilon(\lambda).$$

Then

$$\begin{aligned} 0 &= \lim_{n\to\infty} \left\| A_n\left(e^{-i\lambda}\right) - A\left(e^{-i\lambda}\right) \right\|^2_{L^2(F_\varepsilon)} \\ &= \lim_{n\to\infty} \left\| \tilde{X}_{t,n} - \tilde{X}_t \right\|^2_{L^2(\Omega)}. \end{aligned}$$

This implies

$$\begin{aligned} &\lim_{n\to\infty} \left\| \varphi(B)\,\tilde{X}_{t,n} - \varepsilon_t \right\|^2_{L^2(\Omega)} \\ &= \lim_{n\to\infty} \left\| \int_{-\pi}^{\pi} e^{it\lambda} \left[\varphi\left(e^{-i\lambda}\right) A_n\left(e^{-i\lambda}\right) - \varphi\left(e^{-i\lambda}\right) A\left(e^{-i\lambda}\right)\right] dZ_\varepsilon(\lambda) \right\|^2_{L^2(\Omega)} \\ &= \lim_{n\to\infty} \left\| \varphi(e^{-i\lambda}) \left[A_n(e^{-i\lambda}) - A(e^{-i\lambda})\right] \right\|^2_{L^2(F_\varepsilon)} \\ &= \lim_{n\to\infty} \frac{\sigma_\varepsilon^2}{2\pi} \int_{-\pi}^{\pi} \left|\varphi(e^{-i\lambda})\right|^2 \left|A_n(e^{-i\lambda}) - A(e^{-i\lambda})\right|^2 d\lambda \\ &\le \frac{\sigma_\varepsilon^2}{2\pi} \sup_{\lambda\in[-\pi,\pi]} \left|\varphi(e^{-i\lambda})\right|^2 \lim_{n\to\infty} ||A_n(e^{-i\lambda}) - A(e^{-i\lambda})||^2_{L^2(F_\varepsilon)} = 0 \end{aligned}$$

and hence, as $L^2(\Omega)$-limit,

$$\lim_{n\to\infty} \varphi(B)\,\tilde{X}_{t,n} = \varepsilon_t.$$

On the other hand,

$$\begin{aligned} &\lim_{n\to\infty} \left\| \varphi(B)\,\tilde{X}_{t,n} - \varphi(B)\,\tilde{X}_t \right\|^2_{L^2(\Omega)} \\ &= \lim_{n\to\infty} \left\| \int_{-\pi}^{\pi} e^{it\lambda} \left[\varphi\left(e^{-i\lambda}\right) A_n\left(e^{-i\lambda}\right) - \varphi\left(e^{-i\lambda}\right) A\left(e^{-i\lambda}\right)\right] dZ_\varepsilon(\lambda) \right\|^2_{L^2(\Omega)} \\ &\le \frac{\sigma_\varepsilon^2}{2\pi} \sup_{\lambda\in[-\pi,\pi]} \left|\varphi(e^{-i\lambda})\right|^2 \lim_{n\to\infty} ||A_n(e^{-i\lambda}) - A(e^{-i\lambda})||^2_{L^2(F_\varepsilon)} = 0, \end{aligned}$$

so that

$$\lim_{n\to\infty} \varphi(B)\tilde{X}_{t,n} = \varphi(B)\tilde{X}_t$$

and hence

$$\varphi(B)\tilde{X}_t \underset{L^2(\Omega)}{=} \varepsilon_t \ (t \in \mathbb{Z}).$$

Thus, we define

$$X_t = \tilde{X}_t.$$

Moreover,

$$\sum_{j=-\infty}^{\infty} |a_j| < \infty$$

implies

$$X_t \underset{a.s.}{=} A(B)\varepsilon_t$$

and

$$\varphi(B)X_t \underset{a.s.}{=} \varphi(B)A(B)\varepsilon_t \underset{a.s.}{=} \varepsilon_t.$$

d) Stationarity:

$$X_t = A(B)\varepsilon_t = \sum_{j=-\infty}^{\infty} a_j\varepsilon_{t-j} \Rightarrow X_t \text{ stationary}$$

e) Uniqueness:

$$X_t \underset{a.s.}{=} A(B)\varepsilon_t$$

Let

$$X_t^* \ (t \in \mathbb{Z}) \text{ s.t. } \varphi(B)X_t^* \underset{a.s.}{=} \varepsilon_t.$$

Then

$$X_t^* \underset{a.s.}{=} A(B)\varphi(B)X_t^* \underset{a.s.}{=} A(B)\varepsilon_t \underset{a.s.}{=} X_t.$$

Remark 6.1 For

$$\varphi(z) = 1 - \sum_{j=1}^{p} \varphi_j z^j,$$

$$\exists c, z_1, \ldots, z_p \in \mathbb{C} \setminus \{0\} \text{ s.t. } \varphi(z) = c \prod_{j=1}^{p} \left(1 - z_j^{-1} z\right).$$

If

$$\exists j \in \{1, \ldots, p\} \text{ s.t. } \left|z_j\right| = 1,$$

then

$$\nexists a_j \in \mathbb{C} \text{ s.t.:}$$

$$\varphi(z) = \prod_{|z_j|=1} \left(1 - z_j^{-1} z\right) \prod_{|z_j| \neq 1} \left(1 - z_j^{-1} z\right) = \sum_{j=-\infty}^{\infty} a_j z^j \; (|z| = 1)$$

so that a stationary solution does not exist.

Lemma 6.1

$$\varphi(z) = 0 \Rightarrow \varphi(\bar{z}) = 0$$

Proof

$$\varphi_j \in \mathbb{R} \; (j = 1, \ldots, p)$$

$$\Rightarrow \varphi(\bar{z}) = \overline{\varphi(z)} \Rightarrow \varphi(z) = 0 \text{ implies } \varphi(\bar{z}) = 0$$

6.3 Causal Stationary Solution

Problem 6.2 Existence of a causal stationary solution.

Solution 6.2 Conditions on $\varphi(z)$.

Theorem 6.2 *Assumptions:*

$$\varepsilon_t \; (t \in \mathbb{Z}) \; iid, \; E(\varepsilon_t) = 0, \; \sigma_\varepsilon^2 = var(\varepsilon_t) < \infty$$

$$\nexists z \in \mathbb{C} \; s.t. \; \varphi(z) = \psi(z) = 0.$$

Then

$$\exists! X_t = \sum_{j=0}^{\infty} a_j \varepsilon_{t-j} \text{ causal stationary s.t.}$$

$$\forall t \in \mathbb{Z} : \varphi(B) X_t \underset{a.s.}{=} \psi(B) \varepsilon_t$$

and

$$\sum_{j=0}^{\infty} |a_j| < \infty$$

$$\Leftrightarrow$$

$$\varphi(z) \neq 0 \text{ for } z \text{ with } |z| \leq 1$$

Proof

1) "$\Leftarrow$": w.l.o.g. $\psi \equiv 1$.

$$\varphi(z) \neq 0 \text{ for } z \text{ with } |z| \leq 1$$

$$\Rightarrow X_t = A(B) \varepsilon_t = \frac{1}{\varphi(B)} \varepsilon_t = \text{unique stationary solution}$$

$$\varphi(z) \text{ polynomial}, \varphi(z) \neq 0 \ (|z| \leq 1) \Rightarrow A(z) \ (|z| \leq 1) \text{ holomorphic}$$

$$\Rightarrow \exists \delta > 0 \text{ s.t. } A(z) \ (|z| < r = 1 + \delta) \text{ holomorphic}$$

$$\Rightarrow A(z) = \sum_{j=0}^{\infty} a_j z^j \ (|z| < r) \text{ and } \sum_{j=0}^{\infty} |a_j| < \infty$$

$$\Rightarrow X_t \underset{a.s.}{=} \sum_{j=0}^{\infty} a_j \varepsilon_{t-j}$$

2) "$\Rightarrow$":

$$X_t = A(B) \varepsilon_t = \sum_{j=0}^{\infty} a_j \varepsilon_{t-j} \text{ unique stationary causal solution}$$

and

$$\forall |z| \leq 1 : |A(z)| \leq \sum_{j=0}^{\infty} |a_j| < \infty.$$

Suppose that

$$|z_0| \leq 1 \text{ and } \varphi(z_0) = 0.$$

Then

$$\varphi(B) A(B) \varepsilon_t = \psi(B) \varepsilon_t$$

implies

$$\varphi(z) A(z) = \psi(z) = \sum_{j=0}^{q} \psi_j z^j \ (|z| \leq 1).$$

Since

$$|A(z)| < \infty \ (|z| \leq 1),$$

$$\varphi(z_0) = 0 \text{ implies } \psi(z_0) = 0.$$

However, by assumption,

$$\nexists z \in \mathbb{C} \text{ s.t. } \varphi(z) = \psi(z) = 0.$$

Hence,

$$\nexists z_0 \in \mathbb{C} \text{ s.t. } |z_0| \leq 1 \text{ and } \varphi(z_0) = 0.$$

Remark 6.2

$$\nexists z \in \mathbb{C} \text{ s.t. } \varphi(z) = \psi(z) = 0$$

is a reasonable condition, because otherwise we may get trivial or unnecessarily complicated equations, and possibly infinitely many solutions.

Example 6.1

$$X_t - \varphi X_{t-1} = \varepsilon_t - \varphi \varepsilon_{t-1}$$

$$\Rightarrow X_t = \varepsilon_t \text{ is a solution} \Rightarrow \text{parameter } \varphi \text{ meaningless}$$

In particular,

$$X_t - X_{t-1} = \varepsilon_t - \varepsilon_{t-1}$$

$$\Rightarrow \text{infinitely many solutions } X_t = \mu + \varepsilon_t \ (\mu \in \mathbb{R})$$

6.4 Causal Invertible Stationary Solution

Problem 6.3 Existence of a causal invertible stationary solution.

Solution 6.3 Conditions on $\varphi(z)$ and $\psi(z)$.

Theorem 6.3 *Assumptions:*

$$\varepsilon_t \ (t \in \mathbb{Z}) \ iid, \ E(\varepsilon_t) = 0, \ \sigma_\varepsilon^2 = var(\varepsilon_t) < \infty$$

$$\nexists z \in \mathbb{C} \ s.t. \ \varphi(z) = 0 = \psi(z)$$

Then

$$\exists! X_t \ causal \ invertible \ stationary \ s.t.$$

$$\forall t \in \mathbb{Z} : \varphi(B) X_t \underset{a.s.}{=} \psi(B) \varepsilon_t$$

and

$$\sum_{j=0}^{\infty} |a_j| < \infty$$

$$\Leftrightarrow$$

$$\varphi(z) \neq 0 \ for \ z \ with \ |z| \leq 1$$

$$\psi(z) \neq 0 \ for \ z \ with \ |z| \leq 1$$

Proof Analogous to the proof of Theorem 6.2.

Example 6.2 For

$$X_t - \varphi X_{t-1} = \varepsilon_t - \psi \varepsilon_{t-1},$$

$$\exists! \text{causal invertible stationary solution} \Leftrightarrow |\varphi|, |\psi| < 1$$

Example 6.3 For

$$X_t - \varphi_1 X_{t-1} - \varphi_2 X_{t-2} = \varepsilon_t, \ \varphi_2 \neq 0,$$

we have

$$1 - \varphi_1 z - \varphi_2 z^2 = 0 \Rightarrow z_{1,2} = \frac{-\varphi_1 \pm \sqrt{\varphi_1^2 + 4\varphi_2}}{2\varphi_2}.$$

Therefore a causal invertible stationary solution exists, if

$$|z_{1,2}| > 1 \Leftrightarrow -1 < \varphi_2 < \min\{1 - \varphi_1, 1 + \varphi_1\}$$

6.5 Autocovariances of ARMA Processes

In this section we consider calculation of $\gamma_X(k)$ ($k \in \mathbb{Z}$) of an $ARMA(p,q)$ process and characterize $\gamma_X(k)$ for $k \to \infty$.

6.5.1 Calculation by Integration

From the previous section we obtain:

Lemma 6.2 *Assumption:*

$$X_t = \textit{stationary causal invertible ARMA}\,(p,q)\,\textit{process.}$$

Then

$$\gamma_X(k) = \frac{\sigma_\varepsilon^2}{2\pi}\int_{-\pi}^{\pi} e^{ik\lambda}\left|\frac{\psi\left(e^{-i\lambda}\right)}{\varphi\left(e^{-i\lambda}\right)}\right|^2 d\lambda.$$

Proof Follows from

$$\gamma_X(k) = \int_{-\pi}^{\pi} e^{ik\lambda} f_X(\lambda)\, d\lambda,$$

$$f_X(\lambda) = \frac{\sigma_\varepsilon^2}{2\pi}\left|\frac{\psi\left(e^{-i\lambda}\right)}{\varphi\left(e^{-i\lambda}\right)}\right|^2.$$

Problem 6.4 Numerical integration of $\exp(ik\lambda)f_X(\lambda)$ is difficult for large k.

Solution 6.4 Derive methods that avoid integration.

6.5.2 Calculation Using the Autocovariance Generating Function

Definition 6.2 Let

$$X_t\ (t \in \mathbb{Z}) \text{ weakly stationary},$$

$$D_G = \left\{ z \in \mathbb{C} \text{ s.t. } \left|\sum_{k=-\infty}^{\infty} \gamma_X(k)\, z^k\right| < \infty \right\}.$$

Then

$$G_X(z) = \sum_{k=-\infty}^{\infty} \gamma_X(k) z^k \ (z \in D_G)$$

is called autocovariance generating function.

Remark 6.3 Application to the calculation of $\gamma_X(k)$: If the series expansion

$$G_X(z) = \sum_{k=-\infty}^{\infty} \gamma_X(k) z^k$$

is known, then the coefficients $\gamma_X(k)$ are known.

First we state some elementary results (see e.g. Lang 2003).

Lemma 6.3 *Assumptions:*

$$\alpha_k \in \mathbb{C} \ (k \in \mathbb{Z})$$

$$r_1 = \limsup_{k \to \infty} |\alpha_{-k}|^{1/k} < r_2 = \frac{1}{\limsup_{k \to \infty} |\alpha_k|^{1/k}}$$

Then

$$\alpha(z) = \sum_{k=-\infty}^{\infty} \alpha_k z^k \in \mathbb{C} \Leftrightarrow r_1 < |z| < r_2$$

Corollary 6.1 *Assumptions:*

$$\alpha_k \in \mathbb{C} \ (k \in \mathbb{Z}), \ \sum_{k=-\infty}^{\infty} |\alpha_k| < \infty.$$

Then

$$\exists r \in (1, \infty) \ s.t. \sum_{k=-\infty}^{\infty} \alpha_k z^k \in \mathbb{C} \ (r^{-1} < |z| < r).$$

Corollary 6.2 *Assumptions:*

$$r = \frac{1}{\limsup_{k \to \infty} |\gamma_X(k)|^{1/k}} > 1$$

Then

$$D_G = \left\{z \in \mathbb{C} \; s.t. \; r^{-1} < |z| < r\right\}.$$

Proof

$$r_1 = \limsup_{k \to -\infty} |\gamma_X(k)|^{1/k} = \limsup_{k \to \infty} |\gamma_X(k)|^{1/k} = r^{-1}$$

$$r_2 = \frac{1}{\limsup_{k \to \infty} |\gamma_X(k)|^{1/k}} = r.$$

Then

$$r > 1 \Rightarrow r_1 < r_2$$

so that Lemma 6.3 implies

$$|G_X(z)| < \infty \Leftrightarrow r^{-1} < |z| < r.$$

Example 6.4 For

$$\gamma_X(k) = \varphi^{|k|} \; (0 < |\varphi| < 1),$$

$$r^{-1} = \limsup_{k \to \infty} |\gamma_X(k)|^{1/k} = |\varphi| < 1$$

Example 6.5 For

$$\gamma_X(k) \underset{k \to \infty}{\sim} c_\gamma k^{-\alpha} \; (\alpha > 0),$$

$$r^{-1} = \limsup_{k \to \infty} |\gamma_X(k)|^{1/k} = 1$$

so that Corollary 6.2 is not applicable.

Corollary 6.3 *Assumptions:*

$$\sum_{k=-\infty}^{\infty} |\gamma_X(k)| < \infty$$

Then

$$\exists r > 1 \; s.t. \; D_G = \left\{z \in \mathbb{C} \; s.t. \; r^{-1} < |z| < r\right\}.$$

Proof Follows from Corollary 6.1.

Lemma 6.4 *Assumption:*

$$X_t \in \mathbb{R}\ (t \in \mathbb{Z})\ \textit{weakly stationary}$$

Then

$$\gamma_X(k) = \delta_{0k}\gamma_X(0)$$
$$\Leftrightarrow$$
$$\exists c \in \mathbb{R}_+ \ s.t.\ G_X\left(e^{-i\lambda}\right) \equiv c$$

Proof

1) "$\Rightarrow$"

$$\gamma_X(k) = \delta_{0k}\gamma_X(0) \in \mathbb{R}_+$$
$$\Rightarrow G_X(z) = \gamma_X(0) \equiv c \in \mathbb{R}_+\ (z \in \mathbb{C})$$

2) "$\Leftarrow$"

$$G_X\left(e^{-i\lambda}\right) \equiv c$$
$$\Rightarrow H(\lambda) = G_X\left(e^{-i\lambda}\right) = \sum_{u=-\infty}^{\infty} \gamma_X(u)\, e^{-iu\lambda} \in L^2[-\pi,\pi]$$
$$\Rightarrow \gamma_X(k) = \frac{1}{2\pi}\int_{-\pi}^{\pi} e^{ik\lambda} G\left(e^{-i\lambda}\right) d\lambda = \delta_{0k}c = \delta_{0k}\gamma_X(0)$$

Lemma 6.5 *Assumption:*

$$\sum_{k=-\infty}^{\infty} |\gamma_X(k)| < \infty$$

Then

$$\exists f_X(\lambda)\ \textit{and}\ f_X(\lambda) = \frac{1}{2\pi} G_X\left(e^{-i\lambda}\right)\ (\lambda \in [-\pi,\pi])$$

Proof Follows from Corollary 3.10.

Lemma 6.6 *Assumptions:*

$$\varepsilon_t\ (t \in \mathbb{Z}),\ E(\varepsilon_t) = 0,\ cov(\varepsilon_s, \varepsilon_t) = \delta_{st}\sigma_\varepsilon^2 < \infty,$$

$$\exists r \in (1, \infty)\ s.t. \sum_{j=-\infty}^{\infty} |a_j|\,|z|^j < \infty\ (r^{-1} < |z| < r),$$

$$X_t = A(B)\,\varepsilon_t = \sum_{j=-\infty}^{\infty} a_j \varepsilon_{t-j}\ (t \in \mathbb{Z}).$$

Then

$$G_X(z) = \sigma_\varepsilon^2 A(z) A\left(z^{-1}\right)\ (r^{-1} < |z| < r)$$

Proof

$$\gamma_X(k) = \sigma_\varepsilon^2 \sum_{j=-\infty}^{\infty} a_j a_{j+k}.$$

$$\Rightarrow G_X(z) = \sum_{k=-\infty}^{\infty} \gamma_X(k)\, z^k = \sigma_\varepsilon^2 \sum_{k=-\infty}^{\infty} \left(\sum_{j=-\infty}^{\infty} a_j a_{j+k} \right) z^k$$

$$= \sigma_\varepsilon^2 \sum_{j=-\infty}^{\infty} a_j z^{-j} \left(\sum_{k=-\infty}^{\infty} a_{j+k} z^{j+k} \right) = \sigma_\varepsilon^2 A\left(z^{-1}\right) A(z).$$

Corollary 6.4 *Assumptions:*

$$X_t = \textit{stationary ARMA}\ (p, q)\ \textit{process}.$$

Then

$$\exists r \in (1, \infty)\ s.t.\ G_X(z) = \sigma_\varepsilon^2 \frac{\psi(z)}{\varphi(z)} \frac{\psi\left(z^{-1}\right)}{\varphi\left(z^{-1}\right)}\ (r^{-1} < |z| < r)$$

Proof

$$X_t = \sum_{j=-\infty}^{\infty} a_j \varepsilon_{t-j} = \int_{-\pi}^{\pi} e^{it\lambda} \frac{\psi\left(e^{-i\lambda}\right)}{\varphi\left(e^{-i\lambda}\right)} dZ_\varepsilon(\lambda).$$

with

$$\sum_{j=-\infty}^{\infty} |a_j| < \infty$$

and hence

$$\frac{\psi(z)}{\varphi(z)} = \sum_{j=-\infty}^{\infty} a_j z^j \text{ and } \sum_{j=-\infty}^{\infty} |a_j| \, |z|^j < \infty \; (r^{-1} < |z| < r).$$

Thus Lemma 6.6 applies.

6.5.3 Calculation Using the Wold Representation

Lemma 6.7 *Assumption:*

$$X_t = \text{causal stationary ARMA}\,(p, q) \text{ process},$$

$$A(z) = \frac{\psi(z)}{\varphi(z)}.$$

Then

$$\gamma_X(k) = \sigma_\varepsilon^2 \sum_{j=0}^{\infty} a_j a_{j+|k|} \; (k \in \mathbb{Z})$$

where

$$A(z) = \sum_{j=0}^{\infty} a_j z^j \; (|z| \leq 1).$$

Proof Follows from Theorem 6.2.

Example 6.6 For

$$|\varphi_1| < 1,$$
$$X_t = \varphi_1 X_{t-1} + \varepsilon_t \; (t \in \mathbb{Z}),$$

we have

$$A(z) = \frac{1}{1 - \varphi_1 z} = \sum_{j=0}^{\infty} \varphi_1^j z^j$$

and hence

$$\gamma_X(k) = \sigma_\varepsilon^2 \sum_{j=0}^{\infty} \varphi_1^j \varphi_1^{j+k} = \varphi_1^k \frac{\sigma_\varepsilon^2}{1 - \varphi_1^2} \; (k \geq 0).$$

6.5.4 Recursive Calculation

Lemma 6.8 *Assumption:*

$$X_t = \text{stationary } AR\,(p) \text{ process}$$

Then

$$\gamma_X(0) = \varphi_1\gamma_X(1) + \ldots + \varphi_p\gamma_X(p) + \sigma_\varepsilon^2, \tag{6.1}$$

$$\gamma_X(k) = \varphi_1\gamma_X(k-1) + \ldots + \varphi_p\gamma_X(k-p) \;\; (k \geq 1). \tag{6.2}$$

Proof

$$X_t = \varphi_1 X_{t-1} + \ldots + \varphi_p X_{t-p} + \varepsilon_t$$
$$\Downarrow \cdot X_{t-k} \text{ and } E(.)$$

$$\begin{aligned}\gamma_X(k) &= E(X_t X_{t-k}) \\ &= \varphi_1 E(X_{t-1}X_{t-k}) + \ldots + \varphi_p E\left(X_{t-p}X_{t-k}\right) + E(\varepsilon_t X_{t-k}) \\ &= \varphi_1\gamma_X(k-1) + \ldots + \varphi_p\gamma_X(k-p) + \delta_{0k}\sigma_\varepsilon^2.\end{aligned}$$

Remark 6.4 Equations (6.1), (6.2) can be used to calculate $\gamma_X(k)$ recursively: (1) calculate initial values $\gamma_X(k)$ $(k = 0, \ldots, p-1)$; (2) for $k \geq p$, apply (6.2).

Lemma 6.9 *Assumptions:*

$$X_t = \text{stationary causal } ARMA\,(p,q) \text{ process}$$

$$A(z) = \sum_{j=0}^{\infty} a_j z^j = \frac{\psi(z)}{\varphi(z)}$$

Then

$$\gamma_X(k) = \sum_{j=1}^{p} \varphi_j\gamma_X(k-j) + \sigma_\varepsilon^2 \sum_{j=k}^{q} \psi_j a_{j-k} \;(k = 0, \ldots, \max(p, q+1) - 1),$$

and

$$\gamma_X(k) = \sum_{j=1}^{p} \varphi_j\gamma_X(k-j) \;\; (k \geq \max(p, q+1)).$$

Proof Analogous.

6.5.5 Asymptotic Decay

Theorem 6.4 *Assumptions:*

$$X_t \in \mathbb{R}\ (t \in \mathbb{Z}) \text{ weakly stationary}$$

$$\text{Wold decomposition } X_t = A(B)\,\varepsilon_t = \sum_{j=0}^{\infty} a_j \varepsilon_{t-j}$$

$$\exists 0 < C_1 < \infty, 0 < \beta < 1 \text{ s.t. } |a_j| \le C_1 \beta^j \ (j \in \mathbb{N}).$$

Then

$$\exists 0 < C_2 < \infty \text{ s.t. } |\gamma(k)| \le C_2 \beta^k.$$

Proof

$$|\gamma(k)| \le \sigma_\varepsilon^2 \sum_{j=0}^{\infty} |a_j| |a_{j+k}| \le \sigma_\varepsilon^2 C_1^2 \sum_{j=0}^{\infty} \beta^{2j+k} = C_2 \beta^k.$$

Remark 6.5 Application to ARMA(p, q) processes: w.l.o.g. we may assume $q = 0$. Then

$$X_t = A(B)\,\varepsilon_t, A(z) = \sum_{j=0}^{\infty} a_j z^j = \frac{1}{\varphi(z)}.$$

Problem 6.5 Asymptotic behaviour of a_j?

Solution 6.5 The solution follows from the state space representation.

Notation 15

$$\underset{p\times 1}{Y_t} = \left(X_{t-p+1}, \ldots, X_t\right)^T, \ \underset{p\times 1}{E_t} = (0, \ldots, 0, \varepsilon_t)^T$$

$$\underset{p\times p}{F} = \left(F_{jl}\right)_{j,l=1,\ldots p} = \begin{pmatrix} 0 & 1 & 0 & \cdots & 0 \\ 0 & 0 & \ddots & \ddots & \vdots \\ \vdots & \vdots & \ddots & \ddots & 0 \\ 0 & 0 & \cdots & 0 & 1 \\ \varphi_p & \varphi_{p-1} & \cdots & \cdots & \varphi_1 \end{pmatrix} = \begin{pmatrix} \underset{(p-1)\times 1}{0} & \underset{(p-1)\times(p-1)}{I} \\ \varphi_p & \varphi_{p-1}, \cdots, \varphi_1 \end{pmatrix}$$

Lemma 6.10 *Assumptions:*

$$\varphi(z) = 1 - \sum_{j=1}^{p} \varphi_j z^j,\ \varphi_p \neq 0$$

$$\varphi(B) X_t = \varepsilon_t \ (t \in \mathbb{Z})$$

Then

$$Y_t = FY_{t-1} + E_t \ (t \in \mathbb{Z}). \tag{6.3}$$

Proof Follows directly by comparing the equations.

Remark 6.6 Equation (6.3) is called state space representation of X_t.

Remark 6.7

$$cov(\varepsilon_s, \varepsilon_t) = \sigma_\varepsilon^2 \delta_{st} \Rightarrow cov(E_s, E_t) = \sigma_\varepsilon^2 \delta_{st} \begin{pmatrix} \underset{(p-1)\times(p-1)}{0} & \underset{(p-1)\times 1}{0} \\ \underset{1\times(p-1)}{0} & 1 \end{pmatrix}$$

Problem 6.6

$$Y_t = FY_{t-1} + E_t = \ldots \underset{?}{=} \sum_{j=0}^{\infty} F^j E_{t-j}$$

Solution 6.6 Characterize eigenvalues of F.

Lemma 6.11

$$\det(F) = (-1)^{p-1} \varphi_p$$

Proof

$$\underset{p\times p}{F} = \begin{pmatrix} \underset{(p-1)\times 1}{0} & \underset{(p-1)\times(p-1)}{I} \\ \varphi_p & \varphi_{p-1}, \cdots, \varphi_1 \end{pmatrix}$$

Switch rows to obtain upper triangular matrix

$$F^* = \begin{pmatrix} \varphi_p & \varphi_{p-1}, \cdots, \varphi_1 \\ \underset{(p-1)\times 1}{0} & \underset{(p-1)\times(p-1)}{I} \end{pmatrix}$$

$$\Rightarrow \det F = (-1)^{p-1} \det F^* = (-1)^{p-1} \varphi_p.$$

Corollary 6.5

$$rank\,(F) = p \Leftrightarrow \varphi_p \neq 0$$

Proof

$$rank\,(F) = p \Leftrightarrow \det F = (-1)^{p-1}\,\varphi_p \neq 0 \Leftrightarrow \varphi_p \neq 0$$

Corollary 6.6 *Assumption:*

$$\varphi_p \neq 0.$$

Then

$$\exists v_1, \ldots, v_p \in \mathbb{R}^p \text{ s.t. } \langle v_j, v_l \rangle = \delta_{jl},\, F v_j = \lambda_j v_j,\, \lambda_j \neq 0$$

Proof

$$\varphi_p \neq 0 \Rightarrow rank\,(F) = p$$

Then

$$\exists \text{orthonormal eigenvectors } v_1, \ldots, v_p,$$
$$\text{eigenvalues } \lambda_1, \ldots, \lambda_p \neq 0.$$

Corollary 6.7 *Assumptions:*

$$\varphi_p \neq 0$$
$$P = \left(P_{jl}\right)_{j,l=1,\ldots,p} = \left(v_1, \ldots, v_p\right)$$
$$\Lambda = diag\left(\lambda_1, \ldots, \lambda_p\right).$$

Then

$$F = P \Lambda P^{-1}.$$

Proof

$$\varphi_p \neq 0 \Rightarrow rank\,(F) = p \Rightarrow \exists P^{-1} = P^T$$
$$\Rightarrow P^{-1} F P = P^T \left(\lambda_1 v_1, \ldots, \lambda_p v_p\right) = \Lambda$$
$$\Rightarrow F = P \Lambda P^{-1}$$

Corollary 6.8 *Assumptions:*

$$\varphi_p \neq 0$$
$$P = \left(P_{jl}\right)_{j,l=1,\dots,p} = \left(v_1, \dots, v_p\right)$$
$$\Lambda = diag\left(\lambda_1, \dots, \lambda_p\right).$$

Then

$$F^j = P\Lambda^j P^{-1}$$

Proof

$$F^j = \left(P\Lambda P^{-1}\right)\left(P\Lambda P^{-1}\right)\dots\left(P\Lambda P^{-1}\right) = P\Lambda^j P.$$

Notation 16

$$F(j) = \left[F_{rs}(j)\right]_{r,s=1,\dots,p} = F^j.$$

Corollary 6.9 *Assumptions:*

$$\varphi_p \neq 0$$
$$|\lambda|_{\max} = \max\left\{|\lambda_1|, \dots, \left|\lambda_p\right|\right\}$$

Then

$$\exists 0 < C < \infty \text{ s.t. } \forall j \in \mathbb{N}: \max_{r,s=1,\dots,p} |F_{rs}(j)| \leq C\,|\lambda|^j_{\max}.$$

Proof Follows from Corollary 6.8.

Problem 6.7

$$|\lambda|_{\max} < 1?$$

Solution 6.7 Follows from conditions on $\varphi(z)$.

Lemma 6.12

$$\det(F - \lambda I) = (-1)^p\left(\lambda^p - \sum_{j=1}^{p}\varphi_j\lambda^{p-j}\right).$$

Proof Proof by induction. Let

$$F_p = \underset{p\times p}{F}.$$

Then

$$F_1 - \lambda I = \varphi_1 - \lambda = (-1)^1 (\lambda - \varphi_1)$$

Suppose that

$$\det\left(F_{p-1} - \lambda I\right) = (-1)^{p-1} \left(\lambda^{p-1} - \sum_{j=1}^{p-1} \varphi_j \lambda^{(p-1)-j}\right).$$

For

$$M = \left(m_{ij}\right)_{i,j=1,\ldots,p} = F_p - \lambda I_p = \begin{pmatrix} -\lambda & 1 & 0 & \cdots & \cdots & 0 \\ 0 & -\lambda & \ddots & \ddots & . & \vdots \\ \vdots & 0 & \ddots & \ddots & \ddots & \vdots \\ \vdots & \vdots & \ddots & \ddots & \ddots & 0 \\ 0 & 0 & \cdots & 0 & -\lambda & 1 \\ \varphi_p & \varphi_{p-1} & \cdots & \cdots & \varphi_2 & \varphi_1 - \lambda \end{pmatrix}$$

let

$$\underset{(p-1)\times(p-1)}{A_{ij}} = F_p - \lambda I_p \text{ without row } i \text{ and without column } j.$$

We apply the Laplace expansion

$$\det\left(F_p - \lambda I_p\right) = \sum_{j=1}^{p} (-1)^{1+j} m_{1j} \det\left(A_{1j}\right).$$

Now

$$\det\left(A_{1j}\right) = 0 \; (j \notin \{1,2\}).$$

Moreover,

$$A_{11} = \begin{pmatrix} -\lambda & 1 & 0 & \cdots & 0 \\ 0 & \ddots & \ddots & \ddots & \vdots \\ \vdots & \ddots & \ddots & \ddots & 0 \\ 0 & \cdots & 0 & -\lambda & 1 \\ \varphi_{p-1} & \cdots & \cdots & \varphi_2 & \varphi_1 - \lambda \end{pmatrix}$$

$$= F_{p-1} - \lambda I_{p-1}.$$

$$\Rightarrow \det(A_{11}) = (-1)^{p-1}\left(\lambda^{p-1} - \sum_{j=1}^{p-1}\varphi_j \lambda^{p-1-j}\right).$$

Also, exchanging $p-2$ columns of

$$A_{12} = \begin{pmatrix} 0 & 1 & 0 & \cdots & \cdots & 0 \\ 0 & -\lambda & \ddots & \ddots & & \vdots \\ \vdots & 0 & \ddots & \ddots & \ddots & \vdots \\ \vdots & & 0 & \ddots & \ddots & 0 \\ 0 & \cdots & \cdots & 0 & -\lambda & 1 \\ \varphi_p & \varphi_{p-2} & \cdots & \cdots & \varphi_2 & \varphi_1 - \lambda \end{pmatrix} = \left(\underset{(p-1)\times 1}{a_{\cdot 1}}, \ldots, \underset{(p-1)\times 1}{a_{\cdot p-1}}\right),$$

we obtain the lower triangular matrix

$$\tilde{A}_{12} = \left(\underset{(p-1)\times 1}{a_{\cdot 2}}, \ldots, \underset{(p-1)\times 1}{a_{\cdot p-1}}, \underset{(p-1)\times 1}{a_{\cdot 1}}\right) = \begin{pmatrix} 1 & 0 & \cdots & \cdots & 0 & 0 \\ -\lambda & \ddots & \ddots & & \vdots & 0 \\ 0 & \ddots & \ddots & \ddots & \vdots & \vdots \\ & 0 & \ddots & \ddots & 0 & \vdots \\ \cdots & \cdots & 0 & -\lambda & 1 & 0 \\ \varphi_{p-2} & \cdots & \cdots & \varphi_2 & \varphi_1 - \lambda & \varphi_p \end{pmatrix}.$$

Hence

$$\det(A_{12}) = (-1)^{p-2}\det(\tilde{A}_{12}) = (-1)^{p-2}\varphi_p.$$

Thus

$$\begin{aligned} \det(F_p - \lambda I_p) &= \sum_{j=1}^{p}(-1)^{1+j} m_{1j}\det(A_{1j}) = -\lambda \cdot \det A_{11} - 1 \cdot \det A_{12} \\ &= (-\lambda)(-1)^{p-1}\left(\lambda^{p-1} - \sum_{j=1}^{p-1}\varphi_j\lambda^{p-1-j}\right) - (-1)^{p-2}\varphi_p \\ &= (-1)^p\left(\lambda^p - \sum_{j=1}^{p-1}\varphi_j\lambda^{p-j} - \varphi_p\right) \\ &= (-1)^p\left(\lambda^p - \sum_{j=1}^{p}\varphi_j\lambda^{p-j}\right). \end{aligned}$$

Corollary 6.10 *Assumptions:*

$$\varphi(z) = 1 - \sum_{j=1}^{p} \varphi_j z^j,\ \varphi_p \neq 0,$$

$$\exists X_t \textit{ causal stationary solution of } \varphi(B)\, X_t = \varepsilon_t \ (t \in \mathbb{Z}),$$

$$\lambda_1, \ldots, \lambda_p = \textit{eigenvalues of } F,$$

$$|\lambda|_{\max} = \max\left\{|\lambda_1|, \ldots, |\lambda_p|\right\}.$$

Then

$$|\lambda|_{\max} < 1.$$

Proof

$$\varphi_p \neq 0 \Rightarrow \textit{rank}\,(F) = p$$

$$\Rightarrow \lambda_1, \ldots, \lambda_p \neq 0$$

Hence,

$$\det(F - \lambda I) = 0 \Leftrightarrow (-1)^p \lambda^{-p} \det(F - \lambda I) = 0.$$

Now,

$$(-1)^p \lambda^{-p} \det(F - \lambda I) = (-1)^p \lambda^{-p} \left[(-1)^p \left(\lambda^p - \sum_{j=1}^{p} \varphi_j \lambda^{p-j}\right)\right]$$

$$= 1 - \sum_{j=1}^{p} \varphi_j \lambda^{-j} = \varphi\left(\frac{1}{\lambda}\right)$$

so that

$$\det(F - \lambda I) = 0 \Leftrightarrow \varphi\left(\frac{1}{\lambda}\right) = 0.$$

By assumption

$$\exists X_t \text{ causal stationary solution of } \varphi(B)\, X_t = \varepsilon_t \ (t \in \mathbb{Z}),$$

which implies

$$\nexists z \in \mathbb{C} \text{ s.t. } |z| \leq 1 \text{ and } \varphi(z) = 0,$$

and hence

$$\frac{1}{|\lambda_j|} > 1 \ (j = 1, \ldots, p).$$

Corollary 6.11 *Assumptions:*

$$\varphi(z) = 1 - \sum_{j=1}^{p} \varphi_j z^j, \ \varphi_p \neq 0,$$

$$\exists X_t \text{ causal stationary solution of } \varphi(B) X_t = \varepsilon_t \ (t \in \mathbb{Z}).$$

Then

$$\exists 0 < C < \infty, \ 0 < \beta < 1 \ s.t. \ \forall j \in \mathbb{N} : \max |F_{rs}(j)| \leq C |\lambda|_{\max}^{j} \leq C\beta^j.$$

Proof Follows from

$$\max_{r,s=1,\ldots,p} |F_{rs}(j)| \leq C |\lambda|_{\max}^{j} \text{ and } |\lambda|_{\max} < 1.$$

Corollary 6.12 *Assumptions:*

$$\varphi(z) = 1 - \sum_{j=1}^{p} \varphi_j z^j, \ \varphi_p \neq 0,$$

$$\exists X_t \text{ causal stationary solution of } \varphi(B) X_t = \varepsilon_t \ (t \in \mathbb{Z}).$$

Then

$$X_t = \sum_{j=0}^{\infty} a_j \varepsilon_{t-j} \ s.t.$$

$$\exists 0 < C < \infty, \ 0 < \beta < 1 \ s.t. \ \forall j \in \mathbb{N} : |a_j| \leq C |\lambda|_{\max}^{j} \leq C\beta^j.$$

Proof

$$|a_j| = |F_{pp}(j)| \leq \max_{1 \leq r,s \leq p} |F_{rs}(j)| \leq C |\lambda|_{\max}^{j}.$$

Corollary 6.13 *Suppose that*

$$\varphi(z) = 1 - \sum_{j=1}^{p} \varphi_j z^j, \ \varphi_p \neq 0,$$

$$\exists X_t \text{ causal stationary solution of } \varphi(B) X_t = \varepsilon_t \ (t \in \mathbb{Z}).$$

Then

$$\exists 0 < C_\gamma < \infty, 0 < \beta_\gamma < 1 \ s.t. \ \forall k \in \mathbb{N} : |\gamma_X(k)| \le C_\gamma \, |\lambda|_{\max}^k \le C_\gamma \beta_\gamma^k.$$

Proof Follows from Theorem 6.4 and Corollary 6.12.

6.6 Integrated, Seasonal and Fractional ARMA and ARIMA Processes

6.6.1 Integrated Processes

Definition 6.3 Let $d \in \mathbb{N}$, and $X_t \in \mathbb{R}$ $(t \in \mathbb{N})$ such that

$$Y_t = (1 - B)^d X_t$$

is a causal invertible stationary ARMA(p, q) process. Then X_t is called an integrated ARMA process of order (p, d, q), or an ARIMA(p, d, q) process. More generally, X_t is called integrated of order d, or $I(d)$, if Y_t is a causal weakly stationary process, and $(1 - B)^{d^*} X_t$ is not stationary for any $d^* < d$.

Remark 6.8 Let

$$\varphi(B) = 1 - \sum_{j=1}^{p} \varphi_j B^j, \ \beta_0, \beta_1, \ldots, \beta_d \in \mathbb{R}, \ d \in \mathbb{N}.$$

Then

$$\varphi(B)(1 - B)^d X_t = \psi(B)\varepsilon_t$$

$$\Leftrightarrow$$

$$\varphi(B)\left\{(1 - B)^d \left(X_t + \sum_{j=0}^{d} \beta_j t^j\right) - \beta_d\right\} = \psi(B)\varepsilon_t.$$

Thus, a polynomial trend of order d may be included without changing the ARIMA(p, d, q) equation.

6.6.2 Seasonal ARMA Processes

Definition 6.4 Let

$$T_0 \in \mathbb{N}, T_0 \geq 2, p, q, r, s \in \mathbb{N},$$

$$\varphi(z) = 1 - \sum_{j=1}^{p} \varphi_j z^j, \ \psi(z) = 1 + \sum_{j=1}^{q} \psi_j z^j,$$

$$\alpha(z) = 1 - \sum_{j=1}^{r} \alpha_j z^j, \ \beta(z) = 1 + \sum_{j=1}^{s} \beta_j z^j.$$

Suppose that

$$\varphi(z) \neq 0 \text{ for } z \text{ with } |z| \leq 1,$$

$$\alpha\left(z^{T_0}\right) \neq 0 \text{ for } z \text{ with } |z| \leq 1,$$

and X_t is the causal stationary solution of

$$\varphi(B)\,\alpha\left(B^{T_0}\right) X_t = \psi(B)\,\beta\left(B^{T_0}\right)\varepsilon_t.$$

Then X_t is called a seasonal ARMA process of order $(p, q) \times (r, s)$ with seasonal period T_0, or a SARMA(p, q, r, s) process.

Example 6.7 Let $-1 < \alpha_1 < 1$. Then

$$\left(1 - \alpha_1 B^{12}\right) X_{t-12} = \varepsilon_t$$

defines a SARMA$(0, 0, 1, 0)$ process with $T_0 = 12$. For $\alpha_1 > 0$,

$$f_X(\lambda) = \frac{\sigma_\varepsilon^2}{2\pi} \frac{1}{\left|1 - \alpha_1 e^{-i12\lambda}\right|^2}$$

has local maxima at $\lambda_0 = 0$, and

$$\lambda_j = \frac{2\pi j}{12} \ (j = 1, \ldots, 6).$$

Example 6.8 Let $-1 < \varphi_1, \alpha_1 < 1$. Then

$$(1 - \varphi_1 B)\left(1 - \alpha_1 B^{12}\right) X_{t-12} = \varepsilon_t$$

defines a SARMA$(1, 0, 1, 0)$ process with $T_0 = 12$ and

$$f_X(\lambda) = \frac{\sigma_\varepsilon^2}{2\pi} \frac{1}{\left|1 - \varphi_1 e^{-i\lambda}\right|^2} \frac{1}{\left|1 - \alpha_1 e^{-i12\lambda}\right|^2}.$$

6.6.3 Fractional ARIMA Processes

Notation 17 *For $x_0 \in [-\infty, \infty]$ and two functions*

$$g : \mathbb{R} \to \mathbb{R} \text{ and } h : \mathbb{R} \to \mathbb{R},$$

we write

$$g(x) \underset{x \to x_o}{\sim} h(x),$$

if

$$\lim_{x \to x_o} \frac{g(x)}{h(x)} = 1.$$

Similarly, for two sequences a_n, b_n ($n \in \mathbb{N}$) we write

$$a_n \sim b_n,$$

if

$$\lim_{n \to \infty} \frac{a_n}{b_n} = 1.$$

Definition 6.5 A function

$$L : \mathbb{R} \to \mathbb{R}$$

is called slowly varying in Karamata's sense, if

$$\forall t > 0 : \lim_{x \to \infty} \frac{L(tx)}{L(x)} = 1.$$

Also, L is called slowly varying at zero in Karamata's sense, if $\tilde{L}(x) = L(x^{-1})$ is slowly varying in Karamata's sense.

Definition 6.6 A function

$$L : \mathbb{R} \to \mathbb{R}$$

is called slowly varying in Zygmund's sense, if

$$\exists x_1 \in \mathbb{R}_+ \text{ s.t. } L(x) > 0 \ (x \geq x_1)$$

and

$$\forall\delta > 0, \exists x_2 \in \mathbb{R}_+ \text{ s.t. } x^{\delta}L(x) \text{ and } x^{-\delta}L(x) \ (x \geq x_2) \text{ are monotone functions.}$$

Also, L is called slowly varying at zero in Zygmund's sense, if $\tilde{L}(x) = L(x^{-1})$ is slowly varying in Zygmund's sense.

Definition 6.7 Let

$$X_t \in \mathbb{R}\ (t \in \mathbb{Z}) \text{ weakly stationary}, \exists f_X = F'_X,$$

$$\exists d \in \left(-\frac{1}{2}, \frac{1}{2}\right), \exists L_f \text{ slowly varying at zero s.t.:}$$

$$f_X(\lambda) \underset{\lambda\to 0}{\sim} L_f(\lambda)\,|\lambda|^{-2d}.$$

Then X_t is said to have long-range dependence, or long memory, or strong dependence, if

$$0 < d < \frac{1}{2},$$

intermediate dependence, if

$$d = 0, \ \lim_{\lambda\to 0} L_f(\lambda) = \infty$$

short-range dependence, or short memory, or weak dependence, if

$$d = 0, \exists L_f(0) = \lim_{\lambda\to 0} L_f(\lambda) < \infty,$$

and antipersistence, if

$$-\frac{1}{2} < d < 0.$$

Theorem 6.5 *Assumptions:*

$$X_t \in \mathbb{R}\ (t \in \mathbb{Z}) \textit{ weakly stationary}, \exists f_X = F'_X,$$

$$d \in \left(0, \frac{1}{2}\right),$$

$$L_f \textit{ slowly varying in Zygmund's sense},$$

$$\forall\varepsilon > 0 : L_f(x) \ (x \in (\varepsilon, \pi)) \textit{ of bounded variation},$$

$$f_X(\lambda) \underset{\lambda\to 0}{\sim} L_f(\lambda)\,|\lambda|^{-2d}.$$

Then

$$\gamma_X(k) \underset{k\to\infty}{\sim} L_\gamma(k)\, k^{2d-1}$$

where

$$L_\gamma(k) = c_\gamma L_f\left(k^{-1}\right),\; c_\gamma = 2\Gamma(1-2d)\sin(\pi d)\,.$$

Proof See e.g. Lighthill (1962), Zygmund (1968).

Theorem 6.6 *Assumptions:*

$$X_t \in \mathbb{R}\ (t \in \mathbb{Z}) \text{ weakly stationary,}$$

$$d \in \left(0, \frac{1}{2}\right),$$

$$L_\gamma \text{ slowly varying in Zygmund's sense,}$$

$$\gamma_X(k) \underset{k\to\infty}{\sim} L_\gamma(k)\, k^{2d-1}.$$

Then

$$f_X(\lambda) \underset{\lambda\to 0}{\sim} L_f(\lambda)\, |\lambda|^{-2d}$$

where

$$L_f(\lambda) = c_f L_\gamma\left(|\lambda|^{-1}\right),\; c_f = \pi^{-1}\Gamma(2d)\sin\left(\frac{\pi}{2} - \pi d\right).$$

Proof See e.g. Lighthill (1962), Zygmund (1968).

We recall a few facts about the gamma function (see e.g. Andrews et al. 1999):

Definition 6.8 The gamma function is defined by

$$\Gamma(z) = \int_0^\infty t^{z-1} e^{-t} dt\ (z \in \mathbb{C},\ \mathrm{Re}\, z > 0).$$

Lemma 6.13 *$\Gamma(z)$ ($z \in \mathbb{C}$, $\mathrm{Re}\, z > 0$) is holomorphic.*

Lemma 6.14

$$\Gamma(z+1) = z\Gamma(z) = z!\ (z \in \mathbb{C},\ \mathrm{Re}\, z > 0).$$

Remark 6.9 Lemma 6.14 motivates an extension of $\Gamma(z)$ to $\mathrm{Re}\, z \leq 0$.

Lemma 6.15 *Let*

$$U_{\Gamma} = \mathbb{C} \setminus \{0, -1, -2, \ldots\}.$$

Then $\Gamma(z)$ $(z \in U_{\Gamma})$ *is holomorphic.*

Lemma 6.16 *Let*

$$w \in \{0, -1, -2, \ldots\}.$$

Then

$$\lim_{z \to w} |\Gamma(z)| = \infty.$$

Lemma 6.17 $\Gamma(z)$ $(z \in \mathbb{C})$ *is meromorphic with simple poles at* $w \in \{0, -1, -2, \ldots\}$.

Lemma 6.18

$$\Gamma(z)\,\Gamma(1-z) = \frac{\pi}{\sin \pi z}$$

Lemma 6.19 (Stirling's Formula)

$$\Gamma(x+1) \underset{x \to \infty}{\sim} \sqrt{2\pi} e^{-x} x^{x+1/2} \; (x \in \mathbb{R}_{+}).$$

Problem 6.8 Generalization of differencing operators $(1-B)^d$ $(d \in \mathbb{N})$ to $d \notin \mathbb{N}$.

Solution 6.8 Use

$$(1-z)^d = \sum_{j=0}^{\infty} \binom{d}{j} (-1)^j z^j$$

and the gamma function to express $d!$ and $(d-j)!$.

Definition 6.9 Let U_{Γ} as in Lemma 6.15 and $a, b \in \mathbb{C}$ s.t.:

$$\{a+1, b+1, a-b+1\} \subset U_{\Gamma}.$$

Then

$$\binom{a}{b} = \frac{\Gamma(a+1)}{\Gamma(b+1)\Gamma(a-b+1)}.$$

The following Lemma is well known from analysis:

Lemma 6.20 *For* $-1 < d < 0$,

$$(1-z)^d = \sum_{j=0}^{\infty} \binom{d}{j} (-1)^j z^j \in \mathbb{C} \; (\,|z| \le 1,\, z \ne 1),$$

and, for $d \ge 0$,

$$(1-z)^d = \sum_{j=0}^{\infty} \binom{d}{j} (-1)^j z^j \in \mathbb{C} \; (\,|z| \le 1).$$

Remark 6.10

$$|\Gamma\,(d-j+1)| = \infty \; (j \ge d+1 \in \mathbb{N})$$

implies

$$\binom{d}{j} = \frac{\Gamma\,(d+1)}{\Gamma\,(j+1)\,\Gamma\,(d-j+1)} = 0 \; (j \ge n+1)$$

so that

$$(1-z)^d = \sum_{j=0}^{d} \binom{d}{j} (-1)^j z^j.$$

Problem 6.9

$$(1-z)^{-d} = \sum_{j=0}^{\infty} a_j z^j = \sum_{j=0}^{\infty} \binom{-d}{j} (-1)^j z^j$$

$$(1-z)^{d} = \sum_{j=0}^{\infty} b_j z^j = \sum_{j=0}^{\infty} \binom{d}{j} (-1)^j z^j$$

$$a_j \underset{j\to\infty}{\sim} ?,\; b_j \underset{j\to\infty}{\sim} ?$$

Solution 6.9 Follows from properties of the gamma function and in particular Stirling's formula.

Lemma 6.21 *Assumption:*

$$d > -1.$$

Then

$$\binom{d}{j}(-1)^j \underset{j\to\infty}{\sim} \frac{1}{\Gamma(-d)} j^{-d-1}.$$

Proof

$$\binom{d}{j} = \frac{\Gamma(d+1)}{\Gamma(j+1)\,\Gamma(d-j+1)} = \frac{1}{\Gamma(j+1)} \frac{\Gamma(d+1)}{\Gamma(d-j+1)}.$$

Then

$$\Gamma(d+1) = \Gamma(d)\, d$$

and

$$\Gamma(d-j+1) = \frac{\Gamma(d)}{(d-1)\cdots(d-j+1)}$$

imply

$$\begin{aligned}\binom{d}{j}(-1)^j &= \frac{d\,(d-1)\cdots(d-j+1)}{\Gamma(j+1)}(-1)^j \\ &= \frac{(-d)\,(1-d)\cdots(j-d-1)}{\Gamma(j+1)} \\ &= \frac{\Gamma(-d)\,(-d)\,(1-d)\cdots(j-d-1)}{\Gamma(j+1)\,\Gamma(-d)} \\ &= \frac{\Gamma(j-d)}{\Gamma(j+1)\,\Gamma(-d)}.\end{aligned}$$

Stirling's formula

$$\Gamma(x) \underset{x\to\infty}{\sim} \sqrt{2\pi} e^{-x+1}\,(x-1)^{(x-1)+\frac{1}{2}}$$

implies, for $x_1, x_2 \to \infty$ with $x_1/x_2 \to c$ $(0 < c < \infty)$,

$$\frac{\Gamma(x_1)}{\Gamma(x_2)} \sim \exp(x_2 - x_1) \frac{(x_1-1)^{(x_1-1)+\frac{1}{2}}}{(x_2-1)^{(x_2-1)+\frac{1}{2}}}.$$

For

$$x_1(j) = j-d,\; x_2(j) = j+1,$$

we have

$$x_2 - x_1 = d + 1, \frac{x_1}{x_2} \to 1$$

and hence

$$\frac{\Gamma(j-d)}{\Gamma(j+1)\,\Gamma(-d)} \underset{j\to\infty}{\sim} \frac{1}{\Gamma(-d)} \exp(d+1) \frac{(j-d-1)^{(j-d-1)+\frac{1}{2}}}{j^{j+\frac{1}{2}}}.$$

Finally,

$$\frac{(j-d-1)^{(j-d-1)+\frac{1}{2}}}{j^{j+\frac{1}{2}}} = \left(1-\frac{d+1}{j}\right)^{j}\left(1-\frac{d+1}{j}\right)^{-d-1}\left(1-\frac{d+1}{j}\right)^{\frac{1}{2}} j^{-d-1}$$
$$\underset{j\to\infty}{\sim} \exp\{-(d+1)\}\, j^{-d-1}$$

so that

$$\frac{\Gamma(j-d)}{\Gamma(j+1)\,\Gamma(-d)} \underset{j\to\infty}{\sim} \frac{1}{\Gamma(-d)} j^{-d-1}.$$

Corollary 6.14 *Assumptions:*

$$d \in \left(-\frac{1}{2}, \frac{1}{2}\right),$$

$$\varepsilon_t \in \mathbb{R}\ (t \in \mathbb{Z}),\ E(\varepsilon_t) = 0,\ cov(\varepsilon_s, \varepsilon_t) = \delta_{st}\sigma_\varepsilon^2 < \infty,$$

$$\Delta_d\left(e^{-i\lambda}\right) = \left(1-e^{-i\lambda}\right)^{d},\ \Delta_{-d}\left(e^{-i\lambda}\right) = \left(1-e^{-i\lambda}\right)^{-d}\ (\lambda \in [-\pi, \pi])$$

Then

$$\Delta_d\left(e^{-i\lambda}\right), \Delta_{-d}\left(e^{-i\lambda}\right) \in L^2(F_\varepsilon).$$

Moreover, for $\lambda \neq 0$,

$$\Delta_d\left(e^{-i\lambda}\right) = \sum_{j=0}^{\infty} b_j e^{-ij\lambda}$$

with

$$b_j \underset{j\to\infty}{\sim} \frac{1}{\Gamma(-d)} j^{-d-1},$$

and

$$\Delta_{-d}\left(e^{-i\lambda}\right)=\sum_{j=0}^{\infty}a_je^{-ij\lambda}$$

with

$$a_j \underset{j\to\infty}{\sim} \frac{1}{\Gamma(d)}j^{d-1}.$$

Proof

1)

$$\left|1-e^{-i\lambda}\right|^{2d}=(2-2\cos\lambda)^d \underset{\lambda\to 0}{\sim} \lambda^{2d},$$

$$d\in\left(-\frac{1}{2},\frac{1}{2}\right)\Rightarrow \min\{-2d,2d\}>-1$$

Hence,

$$\max\left\{\left\|\Delta_d\left(e^{-i\lambda}\right)\right\|^2_{L^2(F_\varepsilon)},\left\|\Delta_{-d}\left(e^{-i\lambda}\right)\right\|^2_{L^2(F_\varepsilon)}\right\}<\infty$$

and

$$\Delta_d\left(e^{-i\lambda}\right),\Delta_{-d}\left(e^{-i\lambda}\right)\in L^2\left(F_\varepsilon\right)$$

2) Lemmas 6.20 and 6.21 imply

$$\Delta_d(z)=\sum_{j=0}^{\infty}b_jz^j \text{ and } \Delta_{-d}(z)=\sum_{j=0}^{\infty}a_jz^j\ (|z|\le 1, z\neq 1)$$

with

$$b_j \underset{j\to\infty}{\sim} \frac{1}{\Gamma(-d)}j^{-d-1} \text{ and } a_j \underset{j\to\infty}{\sim} \frac{1}{\Gamma(d)}j^{d-1}.$$

Corollary 6.15 *Under the assumptions of Corollary 6.14,*

$$\Delta_{-d}(B)\,\varepsilon_t\in L^2(\Omega).$$

Proof

$$\Delta_{-d}\left(e^{-i\lambda}\right)\in L^2\left(F_\varepsilon\right)\Rightarrow\Delta_{-d}(B)\,\varepsilon_t\in L^2(\Omega)$$

Corollary 6.16 *Under the assumptions of Corollary 6.14,*

$$X_t = (1 - B)^{-d} \varepsilon_t = \sum_{j=0}^{\infty} a_j \varepsilon_{t-j} = \sum_{j=0}^{\infty} \binom{-d}{j} (-1)^j \varepsilon_{t-j}$$

is the unique purely stochastic causal invertible stationary solution of

$$(1 - B)^d X_t = \varepsilon_t \ (t \in \mathbb{Z}).$$

Proof

1) Due to Corollary 6.15 we may define

$$\tilde{X}_t = \Delta_{-d}(B)\, \varepsilon_t.$$

2) Let

$$\varepsilon_t = \int_{-\pi}^{\pi} e^{it\lambda} dZ_\varepsilon(\lambda).$$

Then

$$\tilde{X}_t = \int_{-\pi}^{\pi} e^{it\lambda} \left(1 - e^{-i\lambda}\right)^{-d} dZ_\varepsilon(\lambda).$$

Now

$$\left(1 - e^{-i\lambda}\right)^d \neq 0 \ (\lambda \neq 0)$$

implies

$$\left(1 - e^{-i\lambda}\right)^d \left(1 - e^{-i\lambda}\right)^{-d} \underset{L^2(F_\varepsilon)}{=} 1$$

so that

$$(1 - B)^d \tilde{X}_t = \int_{-\pi}^{\pi} \left(1 - e^{-i\lambda}\right)^d \left(1 - e^{-i\lambda}\right)^{-d} e^{it\lambda} dZ_\varepsilon(\lambda)$$
$$= \int_{-\pi}^{\pi} e^{it\lambda} dZ_\varepsilon(\lambda) = \varepsilon_t.$$

Thus, we set

$$X_t = \tilde{X}_t.$$

3) Causality and invertibility follows from

$$(1-z)^{-d}=\sum_{j=0}^{\infty}a_j z^j \text{ and } (1-B)^d=\sum_{j=0}^{\infty}b_j z^j \;(\,|z|\le 1, z\neq 1).$$

4) Uniqueness: Let

$$X_t^* \text{ purely stochastic and } (1-B)^d X_t^*=\varepsilon_t.$$

Then

$$F_{X^*} \text{ continuous at } 0$$

implies

$$\left(1-e^{-i\lambda}\right)^{-d}\left(1-e^{-i\lambda}\right)^{d}\underset{L^2(F_{X^*})}{=}1,$$

and hence

$$\begin{aligned}(1-B)^{-d}(1-B)^d X_t^* &= \int_{-\pi}^{\pi} e^{it\lambda}\left(1-e^{-i\lambda}\right)^{-d}\left(1-e^{-i\lambda}\right)^{d} dZ_{X^*}(\lambda)\\ &= \int_{-\pi}^{\pi} e^{it\lambda} dZ_{X^*}(\lambda)=X_t^*\end{aligned}$$

and

$$X_t^*=(1-B)^{-d}\varepsilon_t=X_t.$$

Corollary 6.17 *Assumptions:*

$$d\in\left(-\frac{1}{2},\frac{1}{2}\right)$$

$$\varepsilon_t\ (t\in\mathbb{Z})\ iid,\ E(\varepsilon_t)=0,\ \sigma_\varepsilon^2=var(\varepsilon_t)<\infty$$

$$\varphi(z)=1-\sum_{j=1}^{p}\varphi_j z^j,\ \psi(z)=\sum_{j=0}^{q}\psi_j z^j$$

$$\nexists z \ s.t.\ \varphi(z)=\psi(z)=0$$

$$\varphi(z)\neq 0\ (\,|z|\le 1),\ \psi(z)\neq 0\ (\,|z|\le 1)$$

$$\sum_{j=0}^{\infty}a_j z^j=(1-z)^{-d}\frac{\psi(z)}{\varphi(z)}$$

Then

$$X_t = (1 - B)^{-d} \varphi^{-1}(B) \psi(B) \varepsilon_t = \sum_{j=0}^{\infty} a_j \varepsilon_{t-j}$$

is the unique purely stochastic causal invertible stationary solution of

$$(1 - B)^d \varphi(B) X_t = \psi(B) \varepsilon_t \ (t \in \mathbb{Z}).$$

Proof Follows from Corollary 6.16 and the results for ARMA processes.

Definition 6.10 Let

$$d \in \left(-\frac{1}{2}, \frac{1}{2}\right),$$

and X_t a causal invertible stationary solution of

$$(1 - B)^d \varphi(B) X_t = \psi(B) \varepsilon_t \ (t \in \mathbb{Z}).$$

Then X_t is called a fractional ARIMA process of order (p, q), or FARIMA(p, d, q) process, or ARFIMA(p, d, q) process.

Corollary 6.18 *Under the assumptions of Corollary 6.17,*

$$\exists F_X'(\lambda) = f_X(\lambda) = \frac{\sigma_\varepsilon^2}{2\pi} \left| \frac{\psi(e^{-i\lambda})}{\varphi(e^{-i\lambda})} \right|^2 \left|1 - e^{-i\lambda}\right|^{-2d}$$

Proof

$$X_t = \left(\sum_{j=0}^{\infty} a_j B^j\right) \varepsilon_t, \ \sum_{j=0}^{\infty} a_j z^j = (1 - z)^{-d} \frac{\psi(z)}{\varphi(z)}, f_\varepsilon(\lambda) = \frac{\sigma_\varepsilon^2}{2\pi}$$

$$\Rightarrow f_X(\lambda) = \frac{\sigma_\varepsilon^2}{2\pi} \left| \sum_{j=0}^{\infty} a_j e^{-ij\lambda} \right|^2 = \frac{\sigma_\varepsilon^2}{2\pi} \left| \frac{\psi(e^{-i\lambda})}{\varphi(e^{-i\lambda})} \right|^2 \left|1 - e^{-i\lambda}\right|^{-2d}$$

Corollary 6.19 *Under the assumptions of Corollary 6.17,*

$$f_X(\lambda) \underset{\lambda \to 0}{\sim} c_f |\lambda|^{-2d}$$

and

$$\gamma_X(k) \underset{k \to \infty}{\sim} c_\gamma |k|^{2d-1}$$

where

$$c_f = \frac{\sigma_\varepsilon^2}{2\pi} \left| \frac{\psi(1)}{\varphi(1)} \right|^2, \; c_\gamma = 2c_f \Gamma(1-2d) \sin \pi d.$$

Proof

1)

$$\left|1 - e^{-i\lambda}\right|^{-2d} \underset{\lambda \to 0}{\sim} |\lambda|^{-2d}, \; \lim_{\lambda \to 0} \frac{\sigma_\varepsilon^2}{2\pi} \left| \frac{\psi\left(e^{-i\lambda}\right)}{\varphi\left(e^{-i\lambda}\right)} \right|^2 = \frac{\sigma_\varepsilon^2}{2\pi} \left| \frac{\psi(1)}{\varphi(1)} \right|^2$$

$$\Rightarrow f_X(\lambda) \underset{\lambda \to 0}{\sim} c_f |\lambda|^{-2d}$$

2)

$$\gamma_X(k) \underset{k \to \infty}{\sim} c_\gamma |k|^{2d-1}$$

follows from Theorem 6.5.

Remark 6.11 Let

$$d = -\frac{1}{2}.$$

Then

$$\binom{-d}{j} (-1)^j = \binom{\frac{1}{2}}{j} (-1)^j \underset{j \to \infty}{\sim} \frac{1}{\Gamma(\frac{1}{2})} k^{-\frac{3}{2}}.$$

Hence

$$\left(1 - e^{-i\lambda}\right)^{-d} \in L^2(F_\varepsilon)$$

and

$$X_t = (1-B)^{-d} \varepsilon_t \in L^2(F_\varepsilon).$$

However,

$$\left(1 - e^{-i\lambda}\right)^{d} \notin L^2(F_\varepsilon)$$

so that

$$X_t = (1-B)^{-d} \varepsilon_t = \sum_{j=0}^{\infty} a_j \varepsilon_{t-j} \text{ noninvertible!}$$

Hence, the equation

$$(1-B)^d X_t = \varepsilon_t$$

is not well defined.

Lemma 6.22 *Under the assumptions of Corollary 6.17,*

$$f_X(\lambda) = \frac{1}{2\pi} \sum_{k=-\infty}^{\infty} \gamma_X(k)\, e^{-ik\lambda} \ (\lambda \in [-\pi, \pi] \setminus \{0\})$$

Proof Follows by applying Fourier series theory in $L^1[-\pi, \pi]$ based on generalized functions (see e.g. Lighthill 1962).

Corollary 6.20 *Under the assumptions of Corollary 6.17, X_t has*

$$\textit{long memory} \Leftrightarrow 0 < d < \frac{1}{2},$$

$$\textit{short memory} \Leftrightarrow d = 0,$$

$$\textit{antipersistence} \Leftrightarrow -\frac{1}{2} < d < 0.$$

Proof Follows from Corollary 6.19.

Example 6.9 Let X_t be a FARIMA$(0, d, 0)$ process. Then

$$f_X(\lambda) = \frac{\sigma_\varepsilon^2}{2\pi} \left|1 - e^{-i\lambda}\right|^{-2d} \underset{\lambda \to 0}{\sim} \frac{\sigma_\varepsilon^2}{2\pi} |\lambda|^{-2d},$$

$$\begin{aligned}\gamma_X(k) &= \frac{\sigma_\varepsilon^2}{2\pi} \int_{-\pi}^{\pi} \cos(k\lambda)\,(2 - 2\cos\lambda)^{-d}\, d\lambda \\ &= \sigma_\varepsilon^2 (-1)^k \frac{\Gamma(1-2d)}{\Gamma(k-d+1)\,\Gamma(1-k-d)}\end{aligned}$$

(see e.g. Gradshteyn and Ryzhik 1965, p. 372), and

$$\rho_X(k) \underset{k \to \infty}{\sim} \frac{\Gamma(1-d)}{\Gamma(d)} k^{2d-1}.$$

6.7 Unit Roots, Spurious Correlation, Cointegration

Definition 6.11 Let X_t $(t \in \mathbb{N})$ be a purely stochastic process and

$$d = \min \{r \in \mathbb{N} \mid (1 - B)^r X_t \text{ weakly stationary}\}.$$

Then X_t is called integrated of order d, and we write $X_t \sim I(d)$. In particular, if $X_t \sim I(1)$ and $Y_t = X_t - X_{t-1}$ is an ARMA process, then X_t is said to have a unit root.

Theorem 6.7 *Assumptions:*

$$(\xi_t, \eta_t) \in \mathbb{R}^2 \ (t \in \mathbb{N}) \ iid, \ E(\xi_t) = E(\eta_t) = 0, \ 0 < \sigma_\xi^2, \sigma_\eta^2 < \infty$$

$$(\xi_t)_{t\in\mathbb{N}} \text{ independent of } (\eta_t)_{t\in\mathbb{N}}$$

$$X_t = \sum_{j=1}^{t} \xi_j, \ Y_t = \sum_{j=1}^{t} \eta_j \ (t \in \mathbb{N})$$

Then

$$\hat{\rho} = \frac{\sum (X_t - \bar{x})(Y_t - \bar{y})}{\sqrt{\sum (X_t - \bar{x}) \sum (Y_t - \bar{y})}} \underset{d}{\rightarrow} \zeta$$

where

$$P(\zeta = 0) = 1.$$

More specifically,

$$\zeta = \frac{\int_0^1 B_1(s)B_2(s)ds - \int_0^1 B_1(s)ds \int_0^1 B_2(s)ds}{\left\{\int_0^1 B_1^2(s)\,ds - \left[\int_0^1 B_1(s)ds\right]^2\right\}^{\frac{1}{2}} \left\{\int_0^1 B_2^2(s)\,ds - \left[\int_0^1 B_2(s)ds\right]^2\right\}^{\frac{1}{2}}}$$

where

$$B_1, B_2 = \text{independent standard Brownian motions.}$$

Proof Consider, for instance,

$$S_n = \sum_{t=1}^{n} X_t Y_t, \ \xi_t \sim N(0, 1), \ \eta_t \sim N(0, 1).$$

Then

$$X_t = \sum_{i=1}^{t} \xi_i \underset{d}{=} B_1(t) \underset{d}{=} n^{\frac{1}{2}} B_1\left(\frac{t}{n}\right),$$

$$Y_t = \sum_{i=1}^{t} \eta_i \underset{d}{=} B_2(t) \underset{d}{=} n^{\frac{1}{2}} B_2\left(\frac{t}{n}\right),$$

and hence

$$\frac{1}{n^2} \sum_{t=1}^{n} X_t Y_t \underset{d}{=} \sum_{t=1}^{n} B_1\left(\frac{t}{n}\right) B_2\left(\frac{t}{n}\right) \frac{1}{n} \underset{d}{\rightarrow} \int_0^1 B_1(s) B_2(s)\, ds.$$

The proofs for all other sums in $\hat{\rho}$ are analogous. For general ξ_t, η_t apply Donsker's functional CLT and the continuous mapping theorem. See e.g. Granger and Newbold (1974), Phillips (1986) for a detailed proof.

Remark 6.12

$$X_t = \sum_{j=1}^{t} \xi_j \sim I(1), \; Y_t = \sum_{j=1}^{t} \eta_j \sim I(1)$$

Then

$$(\xi_t)_{t \in \mathbb{N}} \text{ independent of } (\eta_t)_{t \in \mathbb{N}}$$

implies

$$Y_t = \beta_1 X_t + \varepsilon_t \tag{6.4}$$

with

$$\beta_1 = 0, \varepsilon_t = Y_t \sim I(1),$$
$$\rho = corr(X_t, Y_t) = 0,$$

and

$$\forall \alpha_1, \alpha_2 \in \mathbb{R}: \varepsilon_t = \alpha_1 Y_t + \alpha_2 X_t \sim I(1).$$

Theorem 6.7 (and a related result for the least squares estimator $\hat{\beta}_1$ of β_1) implies

$$\hat{\rho} \underset{d}{\nrightarrow} \rho \text{ and } \hat{\beta}_1 \underset{d}{\nrightarrow} \beta_1.$$

Thus, $\hat{\rho}, \hat{\beta}_1$ and related statistics cannot be used for identifying ρ, β_1. However,

$$Y_t - Y_{t-1} = \beta_1 (X_t - X_{t-1}) + (\varepsilon_t - \varepsilon_{t-1})$$

with

$$\varepsilon_t - \varepsilon_{t-1} \sim I(0).$$

Thus, ρ and β_1 may be estimated consistently from $X_t - X_{t-1}$ and $Y_t - Y_{t-1}$.

Remark 6.13 Although model (6.4) with $\beta_1 \neq 0$ and $\varepsilon_t \sim I(1)$ can be defined theoretically, in economic applications one is rather interested in the case $\varepsilon_t \sim I(0)$, i.e. in finding equilibria of the form

$$\varepsilon_t = \alpha_1 Y_t + \alpha_2 X_t \sim I(0).$$

This leads to the definition of cointegration (Engle and Granger 1987).

Definition 6.12 Suppose that

$$X_t = (X_{t,1}, \ldots, X_{t,m})^T \in \mathbb{R}^m \ (t \in \mathbb{N}),$$
$$\exists d \geq 1 \text{ s.t. } X_{t,j} \sim I(d) \ (j = 1, \ldots, m),$$
$$b \in \mathbb{N}, 1 \leq b \leq d.$$

Then

$$X_{t,1}, \ldots, X_{t,m} \sim CI(d, b), \text{ i.e. they are cointegrated of order } (d, b)$$
$$\Leftrightarrow$$
$$\exists \beta \in \mathbb{R}^m \text{ s.t. } \beta^T X_t \sim I(d-b)$$

Remark 6.14 More general definitions of cointegration exist in the literature (see e.g. Dhrymes 1997, Lütkepohl 2006 and references therein).

Remark 6.15 Testing for cointegration of order $(1, 1)$ can be based on estimating $\varepsilon_t = \beta^T X_t$ and testing

$$H_0 : \varepsilon_t \sim I(1) \text{ vs. } H_1 : \varepsilon_t \sim I(0).$$

A standard method is for instance the Dickey-Fuller test (Dickey and Fuller 1979; Said and Dickey 1984).

Remark 6.16 Cointegration is often modelled in an extended multivariate ARMA setting (see e.g. Johansen 1995).

Chapter 7
Generalized Autoregressive Processes

Most proofs in this chapter are more involved and are therefore omitted or simplified. For literature and detailed proofs see e.g. Berkes et al. (2003, 2004), Bollerslev (1986), Bougerol and Picard (1992a,b), Brandt (1986), Breiman (1968), Brockwell and Cline (1985), Caines (1988), Furstenberg and Kesten (1960), Giraitis et al. (2000), Hannan and Kanter (1977), Kazakevičius and Leipus (2002), Kingman (1973), Nelson (1990).

7.1 Definition of Generalized Autoregressive Processes

Notation 18

$$I_m = diag\left(\underbrace{1,\ldots,1}_{m}\right)$$

$$\|x\| = \sqrt{\sum_{i=1}^{m} x_i^2}\ (x \in \mathbb{R}^m)$$

$$\|A\|_{M(m,m,\mathbb{R})} = \|A\|_M = norm\ of\ A \in M(m,m,\mathbb{R})$$

Definition 7.1 Let

$$A_t \in M(m,m,\mathbb{R})\ \ (t \in \mathbb{Z}),\ B_t \in \mathbb{R}^m\ (t \in \mathbb{Z}),$$

$$(A_t, B_t)\ \ (t \in \mathbb{Z})\ \text{strictly stationary ergodic process},$$

$$X_t = A_t X_{t-1} + B_t\ (t \in \mathbb{Z}). \tag{7.1}$$

J. Beran, *Mathematical Foundations of Time Series Analysis*,
https://doi.org/10.1007/978-3-319-74380-6_7

Then

$$X_t \ (t \in \mathbb{Z}) = \text{generalized autoregressive process.}$$

If (A_t, B_t) are iid, then X_t $(t \in \mathbb{Z})$ is called a generalized autoregressive process with iid coefficients.

7.2 Stationary Solution of Generalized Autoregressive Equations

Definition 7.2 Let

$$X_t \in \mathbb{R}^m \ (t \in \mathbb{Z}) \text{ s.t. (7.1) holds.}$$

Then

$$X_t = \text{nonanticipative strictly stationary solution,}$$
$$\Leftrightarrow$$
$$X_t \ (t \in \mathbb{Z}) \text{ strictly stationary,}$$
$$\forall t \in \mathbb{Z} : X_t \text{ independent of } \{(A_s, B_s), s > t\}.$$

Notation 19

$$\log^+ x = (\log x)^+ = \max\{0, \log x\}$$

Definition 7.3 Let

$$A_t \in M(m, m, \mathbb{R}) \ (t \in \mathbb{Z}) \text{ strictly stationary ergodic process,}$$
$$A = (A_t)_{t \in \mathbb{Z}},$$
$$E\left(\log^+ \|A_0\|\right) < \infty.$$

Then

$$\gamma_A = \inf_{n \in \mathbb{N}} E\left(\frac{1}{n+1} \log \|A_0 A_{-1} \cdots A_{-n}\|_M\right)$$
$$= \text{top Lyapunov exponent}$$

Definition 7.4 Let

$$\|x\| = \text{any norm on } \mathbb{R}^m,$$

$$\|C\|_M = \sup_{x\in\mathbb{R}^m\setminus\{0\}} \frac{\|Cx\|}{\|x\|} = \sup_{\|x\|=1} \|Cx\| \;\; (C \in M(m,m,\mathbb{R})).$$

Then

$$\|C\|_M = \text{spectral norm of } C.$$

Definition 7.5 Let

$$C \in M(m,m,\mathbb{R}),$$

$$\lambda_1, \ldots, \lambda_m = \text{eigenvalues of } C.$$

Then

$$\rho_C = \max |\lambda_i| = \text{spectral radius of } C.$$

Lemma 7.1

$$\rho_C \le \|C\|_M$$

Lemma 7.2 *Assumptions:*

$$A_t \in M(m,m,\mathbb{R}) \;\; (t \in \mathbb{Z}) \textit{ strictly stationary ergodic process,}$$

$$A = (A_t)_{t\in\mathbb{Z}}.$$

Then

$$\gamma_A \le E(\log \|A_0\|).$$

Moreover, for $m = 1$,

$$\gamma_A = E(\log \|A_0\|).$$

Lemma 7.3 *Assumption:*

$$A_t \in M(m,m,\mathbb{R}) \;\; (t \in \mathbb{Z}) \textit{ strictly stationary ergodic process,}$$

$$A = (A_t)_{t\in\mathbb{Z}},$$

$$E\left(\log^+ \|A_0\|\right) < \infty.$$

Then

$$\gamma_A \underset{a.s.}{=} \lim_{n\to\infty} \frac{1}{n} \log \|A_0 A_{-1} \cdots A_{-n}\|_M$$

Proof See Furstenberg and Kesten (1960) and Kingman (1973).

Corollary 7.1 *Assumption:*

$$A = (A_t)_{t\in\mathbb{Z}} \textit{ with } A_t = A_0 \; (t \in \mathbb{Z})$$

Then

$$\gamma_A = \log \rho_{A_0}.$$

Definition 7.6 Let

$$A_t \in M(m, m, \mathbb{R}) \;\; (t \in \mathbb{Z}),\, B_t \in \mathbb{R}^m \; (t \in \mathbb{Z}),$$
$$(A_t, B_t) \;\; (t \in \mathbb{Z}) \text{ strictly stationary ergodic process,}$$
$$\mathcal{H}_m = \text{set of affine hyperplanes in } \mathbb{R}^m,$$
$$H \in \mathcal{H}_m,$$
$$O_H = A_0 H + B_0 = \{ y \in \mathbb{R}^m \mid y = A_0 x + B_0, x \in H \}.$$

Then

$$H = \text{invariant under model (7.1)}$$
$$\Leftrightarrow$$
$$P(O_H \subseteq H) = 1.$$

Moreover, let

$$\mathcal{H}_{m,\text{inv}} = \{ H \in \mathcal{H}_m \mid P(O_H \subseteq H) = 1 \}.$$

Then model (7.1) is called irreducible, if

$$H \in \mathcal{H}_m \Leftrightarrow H = \mathbb{R}^m.$$

Remark 7.1 Idea for solution of (7.1):

$$X_t = A_t A_{t-1} \cdots A_{t-n} X_{t-n-1} + \sum_{j=0}^{n} \left(\prod_{i=0}^{j-1} A_{t-i} \right) B_{t-j}$$

$$= \ldots \underset{?}{=} \sum_{j=0}^{\infty} \left(\prod_{i=0}^{j-1} A_{t-i} \right) B_{t-j}$$

where

$$\prod_{i=0}^{-1} A_{t-i} = I_m.$$

Theorem 7.1 *Assumptions:*

$$(A_t, B_t) \ (t \in \mathbb{Z}) \text{ strictly stationary,}$$

$$A = (A_t)_{t \in \mathbb{Z}},$$

$$X_{t,n} = \sum_{j=0}^{n} \left(\prod_{i=0}^{j-1} A_{t-i} \right) B_{t-j},$$

$$E\left(\log^{+} \|A_t\|_M\right) < \infty, E\left(\log^{+} \|B_t\|\right),$$

$$\gamma_A < 0.$$

Then

$$\exists! X_t \in \mathbb{R}^m \ (t \in \mathbb{Z}) \text{ strictly stationary s.t.:}$$

$$X_t = A_t X_{t-1} + B_t \ (t \in \mathbb{Z}).$$

Moreover,

$$X_{t,n} \underset{a.s.}{\rightarrow} X_t = \sum_{j=0}^{\infty} \left(\prod_{i=0}^{j-1} A_{t-i} \right) B_{t-j}.$$

Proof See Brandt (1986).

Remark 7.2 The converse to this theorem holds under additional assumptions only, for instance as stated in the next two Theorems (Bougerol and Picard 1992a).

Theorem 7.2 *Assumptions:*

$$(A_t, B_t)\ (t \in \mathbb{Z})\ iid$$

$$(7.1)\ irreducible$$

$$X_t = \text{strictly stationary nonanticipative solution of } (7.1),$$

$$X_{t,n} = \sum_{j=0}^{n} \left(\prod_{i=0}^{j-1} A_{t-i} \right) B_{t-j}\ (t \in \mathbb{Z}).$$

Then

$$P\left(\lim_{n\to\infty} \prod_{j=0}^{n} A_{-j} = 0 \right) = 1,$$

$$\forall t \in \mathbb{Z} : P\left(\exists \lim_{n\to\infty} X_{t,n} = X_t \in \mathbb{R}^m \right) = 1,$$

$$X_t = \text{unique strictly stationary solution.}$$

Proof See Bougerol and Picard (1992a).

Theorem 7.3 *Assumptions:*

$$(A_t, B_t)\ (t \in \mathbb{Z})\ iid,$$

$$(7.1)\ irreducible,$$

$$E\left(\log^+ \|A_0\|\right) < \infty,\ E\left(\log^+ \|B_0\|\right) < \infty.$$

Then

$$\exists X_t \text{ strictly stationary nonanticipative solution of } (7.1)$$

$$\Leftrightarrow$$

$$\gamma_A < 0$$

Proof See Bougerol and Picard (1992a).

7.3 Definition of VARMA Processes

Definition 7.7 Let

$$m, p, q \in \mathbb{N},$$

$$\Phi_j \in M(m, m, \mathbb{R}) \ \ (j = 0, 1, \ldots, p), \Phi_0 = I_m,$$

$$\Psi_j \in M(m, m, \mathbb{R}) \ \ (j = 0, 1, \ldots), \Psi_0 = I_m,$$

$$\Phi(z) = \Phi_0 - \sum_{j=1}^{p} \Phi_j, \Psi(z) = \sum_{j=0}^{q} \Psi_j z^j,$$

$$\varepsilon_t \in \mathbb{R}^m \ (t \in \mathbb{Z}) \text{ iid.}$$

Then

$$X_t = \text{autoregressive moving average process of order } (p, q)$$
$$= ARMA(p, q) \ \text{ process}$$

$$\Leftrightarrow$$

$$\Phi(B) X_t = \Psi(B) \varepsilon_t \ (t \in \mathbb{Z}). \tag{7.2}$$

Also, if $m \geq 2$, then

$$X_t = \text{multivariate autoregressive moving average process}$$
$$= \text{vector autoregressive moving average process}$$
$$= VARMA(p, q) \ \text{ process}$$

For $q = 0$,

$$VARMA(p, 0) = VAR(p).$$

Remark 7.3 Possible generalization:

$$\varepsilon_t \in \mathbb{R}^l, \Psi_j \in M(m, l, \mathbb{R})$$

Remark 7.4 Frequently used additional conditions:

$$E\left(\|\varepsilon_t\|^2\right) < \infty, E(\varepsilon_t) = 0, var(\varepsilon_t) = \Sigma_\varepsilon \in GL(m, \mathbb{R}) \tag{7.3}$$

Connection to (7.1):

Lemma 7.4 *Let*

$$\begin{aligned} k &= m(p+q), \\ \underset{m(p+q)\times 1}{Y_t} &= \left(X_t^T, \ldots, X_{t-p+1}^T, \varepsilon_t^T, \ldots, \varepsilon_{t-q+1}^T\right)^T, \\ \underset{m\times m(p-1)}{F} &= \left(\Phi_1, \Phi_2, \cdots, \Phi_{p-1}\right) \in M(m, m(p-1), \mathbb{R}), \\ \underset{m\times m(q-1)}{G} &= \left(\Psi_1, \ldots, \Psi_{q-1}\right) \in M(m, m(q-1), \mathbb{R}), \end{aligned}$$

$$A_t = \underset{m(p+q)\times m(p+q)}{A} = \begin{pmatrix} \underset{m\times m(p-1)}{F} & \underset{m\times m}{\Phi_p} & \underset{m\times m(q-1)}{G} & \underset{m\times m}{\Psi_q} \\ I_{m(p-1)} & \underset{m(p-1)\times m}{0} & \underset{m(p-1)\times m(q-1)}{0} & \underset{m(p-1)\times m}{0} \\ \underset{m\times m(p-1)}{0} & \underset{m\times m}{0} & \underset{m\times m(q-1)}{0} & \underset{m\times m}{0} \\ \underset{m(q-1)\times m(p-1)}{0} & \underset{m(q-1)\times m}{0} & I_{m(q-1)} & \underset{m(q-1)\times m}{0} \end{pmatrix}, \tag{7.4}$$

$$\underset{m(p+q)\times 1}{B_t} = \underset{m(p+q)\times m}{B} \underset{m\times 1}{\varepsilon_t} \tag{7.5}$$

with

$$B = \begin{pmatrix} I_m \\ \underset{m(p-1)\times m}{0} \\ I_m \\ \underset{m(q-1)\times m}{0} \end{pmatrix} \in M(m(p+q), m(p+q), \mathbb{R}).$$

Then

$$\Phi(B) X_t = \Psi(B) \varepsilon_t \; (t \in \mathbb{Z})$$

$$\Leftrightarrow$$

$$Y_t = A_t Y_{t-1} + B_t \; (t \in \mathbb{Z}). \tag{7.6}$$

7.4 Stationary Solution of VARMA Equations

Definition 7.8 Let

$$D_j \in M(m, m, \mathbb{R}) \ \ (j = 1, \ldots, p_D)$$

$$D(z) = \sum_{j=0}^{r_0} D_j z^j.$$

Then

$$D(z) = \text{common left divisor of } \Phi(z), \Psi(z),$$

$$\Leftrightarrow$$

$$\exists V_j \ (j = 1, \ldots, r_1), W_j \ (j = 1, \ldots, r_2) \text{ s.t.:}$$

$$\Phi(z) = D(z) V(z) \ \text{ and } \Psi(z) = D(z) W(z)$$

where

$$V(z) = \sum_{j=0}^{r_1} V_j z^j, W(z) = \sum_{j=0}^{r_2} W_j z^j.$$

Definition 7.9 Let

$$\mathscr{D} = \text{set of common left divisors of } \Phi(z), \Psi(z)$$

Then

$$\Phi^{-1}(z) \Psi(z) = \text{irreducible},$$

$$\Leftrightarrow$$

$$\forall D(z) \in \mathscr{D} : \det D(z) \ \text{ does not depend on } z$$

Definition 7.10

$$X_t = \text{nonanticipative strictly stationary solution of (7.2)}$$

$$\Leftrightarrow$$

$$X_t \text{ strictly stationary,}$$

$$\forall t \in \mathbb{Z} : X_t \text{ independent of } \{\varepsilon_s, s \geq t+1\}.$$

Theorem 7.4 *Assumptions:*

$$\Phi^{-1}(z)\,\Psi(z) \text{ irreducible},$$
$$\forall H \in \mathscr{H}_m : P(\varepsilon_t \in H) < 1,$$
$$E\left(\log^{+}\|\varepsilon_t\|\right) < \infty.$$

Then

$$\exists X_t \text{ strictly stationary nonanticipative solution of (7.2)}$$
$$\Leftrightarrow$$
$$\det \Phi(z) \neq 0 \;(\,|z| \leq 1)$$

Proof Follows from Theorem 7.3. For a detailed proof see Bougerol and Picard (1992a) and Brandt (1986). For special cases also see e.g. Brockwell and Cline (1985), Caines (1988), Hannan and Kanter (1977). In particular note that

$$Y_t = A_t Y_{t-1} + B_t$$

with

$$A_t = A_0,\; B_t = B\varepsilon_t$$

as in (7.4) and (7.5) of Lemma 7.4,

$$rank\,(A_0) = k = m\,(p+q)\,,$$
$$\text{eigenvalues } \lambda_1, \ldots, \lambda_k, \text{ eigenvectors } v_1, \ldots, v_k.$$

Let

$$A = (A_t)_{t\in\mathbb{Z}}\,,$$
$$P = (v_1, \ldots, v_k)\,,\; \Lambda = \operatorname{diag}(\lambda_1, \ldots, \lambda_k)\,.$$

Then

$$\rho_{A_0} = \max_{1\leq j\leq k} \left|\lambda_j\right|,\; \gamma_A = \log \rho_{A_0},$$
$$A_0^n = P\Lambda^n P^{-1}$$

and hence

$$\lim_{n\to\infty} A_0^n = 0 \Leftrightarrow \rho_{A_0} < 1 \Leftrightarrow \gamma_A < 0.$$

Moreover, as in the univariate case,

$$\det(A_0 - \lambda I) = (-1)^{m(p+q)} \lambda^{m(p+q)} \Phi\left(\lambda^{-1}\right).$$

Hence Theorem 7.3 together with

$$\det \Phi(z) \neq 0 \ (|z| \leq 1) \Rightarrow \rho_{A_0} < 1 \Rightarrow \gamma_A = \log \rho_{A_0} < 0$$

implies:

$$\exists X_t \text{ strictly stationary nonanticipative solution of (7.2).}$$

The opposite direction is more involved (see Bougerol and Picard 1992a).

Remark 7.5 Theorem 7.4 is even applicable when $E(||\varepsilon_t||^2) = \infty$.

7.5 Definition of GARCH Processes

Definition 7.11 Let

$$\varepsilon_t \in \mathbb{R} \ (t \in \mathbb{Z}) \text{ iid},$$
$$\sigma_\varepsilon^2 = E\left(\varepsilon_t^2\right) = 1,$$
$$\alpha_0 > 0, \alpha_j \geq 0 \ (j = 1, \ldots, p),$$
$$\beta_j \geq 0 \ (j = 1, \ldots, q).$$

Then

$$X_t = \text{GARCH}(p, q) \text{ process of order } (p, q)$$

$$\Leftrightarrow$$

$$X_t = \sigma_t \varepsilon_t \ (t \in \mathbb{Z}) \tag{7.7}$$

with

$$\sigma_t^2 = \alpha_0 + \sum_{j=1}^{p} \alpha_j X_{t-j}^2 + \sum_{j=1}^{q} \beta_j \sigma_{t-j}^2 \ (t \in \mathbb{Z}). \tag{7.8}$$

In particular, a GARCH$(p, 0)$ process is also called an ARCH(p) process.

7.6 Stationary Solution of GARCH Equations

The existence of a stationary solution of (7.7), (7.8) is possible, even if ε_t are not iid:

Theorem 7.5 *Assumptions:*

$$\varepsilon_t \in \mathbb{R}\ (t \in \mathbb{Z})\ \text{strictly stationary ergodic}$$

$$\sigma_\varepsilon^2 = var(\varepsilon_t) = 1$$

$$p = q = 1$$

$$\alpha_0 > 0,\ \alpha_1 \geq 0,\ \beta_1 \geq 0$$

$$E\left[\left|\log\left(\alpha_1\varepsilon_0^2 + \beta_1\right)\right|\right] < \infty,\ E\left[\log\left(\alpha_1\varepsilon_0^2 + \beta_1\right)\right] < 0$$

Then

$$\exists! X_t\ (t \in \mathbb{Z})\ \text{s.t. (7.7), (7.8) hold}$$

and

$$X_t\ (t \in \mathbb{Z}) = \text{strictly stationary and ergodic process.}$$

Proof For a detailed proof see Bougerol and Picard (1992a,b); also see Berkes et al. (2004). Note that, if a solution exists, then

$$\sigma_t^2 = \alpha_0 + \alpha_1 X_{t-1}^2 + \beta_1\sigma_{t-1}^2$$

$$\begin{aligned}\sigma_t^2 &= \alpha_0 + \alpha_1 X_{t-1}^2 + \beta_1\sigma_{t-1}^2\\ &= \alpha_0 + \left(\alpha_1\varepsilon_{t-1}^2 + \beta_1\right)\sigma_{t-1}^2\\ &= \alpha_0 + \left(\alpha_1\varepsilon_{t-1}^2 + \beta_1\right)\left(\alpha_0 + \alpha_1 X_{t-2}^2 + \beta_1\sigma_{t-2}^2\right)\\ &= \ldots \underset{?}{=} \alpha_0\left[1 + \sum_{l=1}^{\infty}\prod_{j=1}^{l}\left(\alpha_1\varepsilon_{t-j}^2 + \beta_1\right)\right].\end{aligned}$$

For the last equation, we need to prove

$$P\left(\sum_{l=1}^{\infty}\eta_l < \infty\right) = 1$$

where

$$\eta_l = \prod_{j=1}^{l} \left(\alpha_1 \varepsilon_{t-j}^2 + \beta_1\right).$$

Let

$$\kappa = E\left[\log\left(\alpha_1 \varepsilon_0^2 + \beta_1\right)\right] < 0.$$

Then

$$P\left(\lim_{l\to\infty} \frac{1}{l} \log \eta_l = \kappa\right) = 1$$

which then implies

$$P\left(\sum_{l=1}^{\infty} \eta_l < \infty\right) = 1.$$

Theorem 3.5.8 in Stout (1974) implies strict stationarity and ergodicity.

For the GARCH(1, 1) process with iid ε_t the following equivalence can be shown:

Theorem 7.6 *Assumptions:*

$$\varepsilon_t \in \mathbb{R}\ (t \in \mathbb{Z})\ iid$$

$$\sigma_\varepsilon^2 = var\left(\varepsilon_t\right) = 1$$

$$p = q = 1$$

$$\alpha_0 > 0,\ \alpha_1 \geq 0,\ \beta_1 \geq 0$$

$$E\left[\left|\log\left(\alpha_1 \varepsilon_0^2 + \beta_1\right)\right|\right] < \infty$$

Then

$$\exists! X_t\ (t \in \mathbb{Z})\ s.t.\ (7.7), (7.8)\ hold$$

$$\Leftrightarrow$$

$$E\left[\log\left(\alpha_1 \varepsilon_0^2 + \beta_1\right)\right] < 0$$

Moreover,

$$X_t\ (t \in \mathbb{Z}) = strictly\ stationary\ ergodic\ process.$$

Proof For a detailed proof see Nelson (1990), or Bougerol and Picard (1992a,b).

Remark 7.6 More generally, w.l.o.g. $p, q \geq 3$ (some coefficients may be zero). Define

$$k = p + q - 1,$$

$$\underset{1\times(q-1)}{b_t} = \left(\beta_1 + \alpha_1 \varepsilon_{t-1}^2, \beta_2, \ldots, \beta_{q-1}\right),$$

$$\underset{1\times(q-1)}{e_t} = \left(\varepsilon_{t-1}^2, 0, \ldots, 0\right),$$

$$\underset{1\times(p-2)}{a} = \left(\alpha_2, \ldots, \alpha_{p-1}\right)$$

$$\underset{k\times k}{A_t} = \begin{pmatrix} b_t & \beta_q & a & \alpha_p \\ I_{q-1} & 0 & 0 & 0 \\ e_t & 0 & 0 & 0 \\ 0 & 0 & I_{p-2} & 0 \end{pmatrix},$$

$$B_t = B = (\alpha_0, 0, \ldots, 0)^T \in \mathbb{R}^k,$$

$$Y_t = \left(\sigma_t^2, \ldots, \sigma_{t-q+1}^2, X_{t-1}^2, \ldots, X_{t-p+1}^2\right)^T \in \mathbb{R}^k.$$

Then (7.7), (7.8) are equivalent to

$$Y_t = A_t Y_{t-1} + B_t \; (t \in \mathbb{Z}).$$

Theorem 7.7 *Assumptions:*

$$\varepsilon_t \in \mathbb{R} \; (t \in \mathbb{Z}) \text{ strictly stationary ergodic}$$

$$\sigma_\varepsilon^2 = var(\varepsilon_t) = 1$$

$$\alpha_0 > 0, \; \alpha_j \geq 0 \; (j = 1, \ldots, p)$$

$$\beta_j \geq 0 \; (j = 1, \ldots, q)$$

$$E(\log \|A_0\|) < \infty$$

$$\gamma_A < 0$$

Then

$$\exists! X_t \; (t \in \mathbb{Z}) \text{ s.t. (7.7), (7.8) hold.}$$

Moreover

$$X_t \; (t \in \mathbb{Z}) \text{ strictly stationary, ergodic process.}$$

Proof For a detailed proof see Brandt (1986) and Bougerol and Picard (1992a,b). In particular, let

$$X_t\ (t \in \mathbb{Z}) \text{ s.t. } (7.7), (7.8) \text{ hold.}$$

For the GARCH(1, 1) with iid ε_t, the following equivalence statements can be proved:

Theorem 7.8 *Assumptions:*

$$\varepsilon_t \in \mathbb{R}\ (t \in \mathbb{Z}) \text{ iid},$$
$$\sigma_\varepsilon^2 = var(\varepsilon_t) = 1,$$
$$\alpha_0 > 0,\ \alpha_j \geq 0\ (j = 1, \ldots, p),$$
$$\beta_j \geq 0\ (j = 1, \ldots, q),$$
$$E(\log \|A_0\|) < \infty.$$

Then

$$\exists X_t\ (t \in \mathbb{Z}) \text{ s.t. } (7.7), (7.7) \text{ hold.}$$
$$\Leftrightarrow$$
$$\gamma_A < 0$$

Moreover,

$$X_t\ (t \in \mathbb{Z}) \text{ is unique, strictly stationary and ergodic.}$$

Proof See Bougerol and Picard (1992a).

Theorem 7.9 *Assumptions:*

$$\varepsilon_t \in \mathbb{R}\ (t \in \mathbb{Z}) \text{ iid},$$
$$\sigma_\varepsilon^2 = var(\varepsilon_t) = 1,$$
$$\alpha_0 > 0,\ \alpha_j \geq 0\ (j = 1, \ldots, p),$$
$$\beta_j \geq 0\ (j = 1, \ldots, q),$$
$$E(\log \|A_0\|) < \infty,$$
$$X_t\ (t \in \mathbb{Z}) \text{ solution of } (7.7), (7.8).$$

Then

$$\sigma_X^2 = var(X_t) < \infty$$
$$\Leftrightarrow$$
$$\sum_{j=1}^{p} \alpha_j + \sum_{j=1}^{q} \beta_j < 1.$$

Proof For a detailed proof see Bollerslev (1986). Note that stationarity, (7.7), (7.8) and $\sigma_X^2 < \infty$ imply

$$\begin{aligned}\sigma_X^2 &= E(X_t^2) = E(\sigma_t^2) \\ &= \alpha_0 + \sum_{j=1}^{p} \alpha_j E(X_{t-j}^2) + \sum_{j=1}^{q} \beta_j E(\sigma_{t-j}^2) \\ &= \alpha_0 + \sigma_X^2 \left(\sum_{j=1}^{p} \alpha_j + \sum_{j=1}^{q} \beta_j \right).\end{aligned}$$

Hence

$$0 < \alpha_0 = \sigma_X^2 \left[1 - \left(\sum_{j=1}^{p} \alpha_j + \sum_{j=1}^{q} \beta_j \right) \right]$$

and therefore

$$\sum_{j=1}^{p} \alpha_j + \sum_{j=1}^{q} \beta_j < 1,$$

which proofs "$\Rightarrow$".

Theorem 7.10 *Assumptions:*

$$\varepsilon_t \in \mathbb{R} \ (t \in \mathbb{Z}) \text{ strictly stationary ergodic,}$$
$$\sigma_\varepsilon^2 = var(\varepsilon_t) = 1,$$
$$\alpha_0 > 0, \ \alpha_j \geq 0 \ (j = 1, \ldots, p),$$
$$\beta_j \geq 0 \ (j = 1, \ldots, q),$$
$$E(\log \|A_0\|) < \infty,$$
$$\gamma_A < 0.$$

Then

$$\sum_{j=1}^{q} \beta_j < 1.$$

Proof See Bougerol and Picard (1992b).

Lemma 7.5 *Assumptions:*

$$\beta_j \geq 0 \; (j = 1, \ldots, q).$$

$$\beta(z) = 1 - \sum_{j=1}^{q} \beta_j z^j \; (z \in \mathbb{C}).$$

Then

$$\sum_{j=1}^{q} \beta_j < 1$$

$$\Leftrightarrow$$

$$\beta(z) \neq 0 \; (|z| \leq 1)$$

Proof See Berkes et al. (2003).

7.7 Definition of ARCH(∞) Processes

Definition 7.12 Let

$$\xi_t \in \mathbb{R} \; (t \in \mathbb{Z}) \text{ iid}, P(\xi_t > 0) = 1,$$

$$b_j \geq 0 \; (j = 0, 1, 2, \ldots).$$

Then

$$\eta_t = \text{ARCH}(\infty) \text{ process},$$

$$\Leftrightarrow$$

$$\eta_t = \sigma_t^2 \xi_t \; (t \in \mathbb{Z}) \tag{7.9}$$

with

$$\sigma_t^2 = b_0 + \sum_{j=1}^{\infty} b_j \eta_{t-j} \; (t \in \mathbb{Z}). \tag{7.10}$$

7.8 Stationary Solution of ARCH(∞) Equations

Connection to GARCH processes:

Theorem 7.11 *Assumptions:*

$$X_t\ (t \in \mathbb{Z}) \text{ strictly stationary s.t. (7.7), (7.8) hold,}$$

$$\xi_t = \varepsilon_t^2,\ \eta_t = X_t^2,$$

$$\alpha(z) = \sum_{j=1}^{q} \alpha_j z^j,\ \beta(z) = 1 - \sum_{j=1}^{q} \beta_j z^j\ (z \in \mathbb{C}),$$

$$b_0 = \frac{\alpha_0}{\beta(1)},\ \sum_{j=1}^{\infty} b_j z^j = \frac{\alpha(z)}{\beta(z)}\ (z \in \mathbb{C}).$$

Then

$$\eta_t = \sigma_t^2 \xi_t\ (t \in \mathbb{Z})$$

with

$$\sigma_t^2 = b_0 + \sum_{j=1}^{\infty} b_j \eta_{t-j}\ (t \in \mathbb{Z}). \tag{7.11}$$

Proof For a detailed proof see Kazakevičius and Leipus (2002). In particular,

$$\sigma_t^2 \xi_t = \left[\alpha_0 + \alpha(B)\eta_t + (1 - \beta(B))\sigma_t^2\right]\xi_t$$

so that

$$\sigma_t^2 \beta(B) = \alpha_0 + \alpha(B)\eta_t. \tag{7.12}$$

Theorem 7.10 implies

$$\sum_{j=1}^{p} \beta_j < 1$$

so that

$$\beta(1) > 0,\ \beta(z) \neq 0\ (|z| \leq 1).$$

Therefore

$$\frac{\alpha(z)}{\beta(z)} = \sum_{j=1}^{\infty} b_j z^j \in \mathbb{C}\ (|z| \leq 1),$$

and

$$0 < b_0 = \frac{\alpha_0}{\beta(1)} < \infty,\ 0 \leq \sum_{j=1}^{\infty} b_j < \infty.$$

It then can be shown that (7.12) may be rewritten as

$$\begin{aligned}\sigma_t^2 &= \frac{\alpha_0}{\beta(B)} + \frac{\alpha(B)}{\beta(B)}\eta_t \\ &= \frac{\alpha_0}{\beta(1)} + \frac{\alpha(B)}{\beta(B)}\eta_t \\ &= b_0 + \sum_{j=1}^{\infty} b_j \eta_{t-j}.\end{aligned}$$

Sufficient condition for the existence of a stationary ARCH(∞) process:

Theorem 7.12 *Assumptions:*

$$\xi_t \in \mathbb{R}\ (t \in \mathbb{Z})\ iid,\ P(\xi_t > 0) = 1,$$

$$b_j \geq 0\ (j = 0, 1, 2, \ldots),$$

$$E(\xi_t) \sum_{j=1}^{\infty} b_j < 1,$$

$$\eta_t = \xi_t b_0 \sum_{l=0}^{\infty} \sum_{i_1,\ldots,i_l=1}^{\infty} b_{i_1} \cdots b_{i_l} \xi_{t-i_1} \cdots \xi_{t-i_1-\ldots-i_l}\ (t \in \mathbb{Z}).$$

Then

$$\eta_t = \text{unique strictly stationary nonanticipative solution of (7.9), (7.10)},$$

and

$$E(\eta_t) < \infty.$$

Moreover, if

$$\sqrt{E\left(\xi_t^2\right)}\sum_{j=1}^{\infty} b_j < 1,$$

then

$$\eta_t = \textit{weakly stationary solution.}$$

Proof See Giraitis et al. (2000).

Chapter 8
Prediction

8.1 Best Linear Prediction Given an Infinite Past

Problem 8.1

$$X_t \text{ weakly stationary}$$

$$X_t = \mu_t + \sum_{j=0}^{\infty} a_j \varepsilon_{t-j} = \text{Wold decomposition}$$

Find the optimal linear prediction $\hat{X}_{t+k}$ of X_{t+k} given X_s $(s \leq t)$.

Solution 8.1 Lemma 2.25 implies

$$\hat{X}_{t+k} = \mu_{t+k} + \sum_{j=0}^{\infty} a_{j+k}\varepsilon_{t-j} \ (k \geq 1).$$

If $A(B) = \sum_{j=0}^{\infty} a_j B^j$ is invertible with

$$\varepsilon_t = \sum_{j=0}^{\infty} b_j X_{t-j},$$

then Corollary 2.14 implies

$$\hat{X}_{t+k} = \mu + \sum_{j=0}^{\infty} \beta_j (k) X_{t-j}$$

J. Beran, *Mathematical Foundations of Time Series Analysis*,
https://doi.org/10.1007/978-3-319-74380-6_8

with

$$\beta_j(k) = \sum_{l=0}^{j} a_{l+k} b_{j-l} \; (k \in \mathbb{N}).$$

Problem 8.2 Find a prediction interval $I \subseteq \mathbb{R}$ based on $\hat{X}_{t+k}$ s.t.

$$P(X_{t+k} \in I \mid X_s, s \le t) = 1 - \alpha.$$

Solution 8.2 In general, prediction intervals depend on the conditional distribution of

$$e_{t+k} = X_{t+k} - \hat{X}_{t+k} = \sum_{j=0}^{k-1} a_j \varepsilon_{t+k-j}.$$

Notation 20

$$\begin{aligned} F_k(x \mid X_s, s \le t) &= P(e_{t+k} \le x \mid X_s, s \le t) \; (x \in \mathbb{R}) \\ q_k(u \mid X_s, s \le t) &= F_k^{-1}(u \mid X_s, s \le t) \; (u \in [0,1]) \\ F_k(x) &= P(e_{t+k} \le x) = P(e_k \le x) \; (x \in \mathbb{R}) \\ q_k(u) &= F_k^{-1}(u) \; (u \in [0,1]) \end{aligned}$$

Theorem 8.1 *Assumptions:*

$$X_t \; (t \in \mathbb{Z}) \text{ weakly stationary},$$

$$\text{Wold decomposition } X_t = \mu_t + \sum_{j=0}^{\infty} a_j \varepsilon_{t-j},$$

$$k \in \{1, 2, \ldots\}, \alpha \in \left(0, \frac{1}{2}\right),$$

$$a = \hat{X}_{t+k} + q_k\left(\frac{\alpha}{2} \mid X_s, s \le t\right), \; b = \hat{X}_{t+k} + q_k\left(1 - \frac{\alpha}{2} \mid X_s, s \le t\right).$$

Then,

$$P(X_{t+k} \in (a, b] \mid X_s, s \le t) \underset{a.s.}{=} 1 - \alpha.$$

Proof Follows from the definition of q_k.

Corollary 8.1 *Assumptions:*

$$X_t = \sum_{j=0}^{\infty} a_j \varepsilon_{t-j} \ (t \in \mathbb{Z}) \text{ linear process in } L^2(\Omega)$$

Then

$$q_k(u \mid X_s, s \le t) \underset{a.s.}{=} q_k(u) \ \ (u \in [0,1]).$$

Proof Follows from independence of ε_t $(t \in \mathbb{Z})$.

8.2 Predictability

Problem 8.3 Conditions s.t. a weakly stationary process X_t is purely stochastic.

Solution 8.3 The conditions can be derived using results from complex analysis.

Definition 8.1 For complex valued functions $g(z)$ define the conditions

$$g \text{ holomorphic on } \{|z| < 1\} \tag{H1}$$

$$\sup_{0 \le r < 1} \int_{-\pi}^{\pi} \left| g\left(re^{i\lambda}\right) \right|^2 d\lambda < \infty \tag{H2}$$

Then

$$H^2 = \{g \mid \text{conditions (H1) and (H2) hold}\}$$

is called a Hardy space or Hardy class.

Lemma 8.1 *Assumptions:*

$$g \in H^2$$

$$\bar{g}\left(e^{i\lambda}\right) = \lim_{r \uparrow 1} g\left(re^{i\lambda}\right) \not\equiv 0$$

Then

$$\int_{-\pi}^{\pi} \log \left| \bar{g}\left(e^{-i\lambda}\right) \right| d\lambda > -\infty$$

Proof See e.g. Duren (1970), Katznelson (1976), Koosis (1980).

Lemma 8.2 *Assumptions:*

$$f = \textit{spectral density}$$

$$\int_{-\pi}^{\pi} \log f(\lambda)\, d\lambda > -\infty$$

Then

$$\exists g \in H^2 \textit{ s.t.:}$$

$$f(\lambda) = \frac{1}{2\pi} \left| \bar{g}\left(e^{-i\lambda}\right) \right|^2 \; (\lambda_{Lebesgue} - a.s.)$$

Proof See e.g. Duren (1970), Katznelson (1976), Koosis (1980).

Theorem 8.2 *Assumptions:*

$$X_t \textit{ weakly stationary}$$

$$F_X = \textit{spectral distribution function of } X_t$$

$$\nexists c \in \mathbb{R} \textit{ s.t. } P(X_t = c, \forall t \in \mathbb{Z}) = 1$$

Then

$$X_t \textit{ purely stochastic}$$

$$\Leftrightarrow$$

$$F_X << \lambda_{Lebesgue}, \tag{8.1}$$

$$\lambda_{Lebesgue}\text{-}a.s. \; \exists f_X(\lambda) = F_X'(\lambda) > 0, \tag{8.2}$$

$$\int_{-\pi}^{\pi} \log f_X(\lambda)\, d\lambda > -\infty. \tag{8.3}$$

Proof For a detailed proof see e.g. Doob (1953, Theorem 4.3, Ch. XII), (also see e.g. Shiryaev 1995, Ch. VI).

$$E(X_t) = 0 \text{ w.l.o.g.}$$

Then

$$\text{Wold decomposition } X_t = \mu_t + \sum_{j=0}^{\infty} a_j \varepsilon_{t-j},\; E(\mu_t) = 0.$$

Hence,

$$X_t \text{ purely stochastic} \Leftrightarrow \mu_t \equiv 0$$
$$\Leftrightarrow$$
$$F_X << \lambda_{\text{Lebesgue}}$$

and

$$f_X(\lambda) = \frac{\sigma_\varepsilon^2}{2\pi} \left|A\left(e^{-i\lambda}\right)\right|^2 = \frac{\sigma_\varepsilon^2}{2\pi} \left|\sum_{j=0}^{\infty} a_j e^{-ij\lambda}\right|^2.$$

Is this equivalent to (8.1)–(8.3)?

1) "$\Rightarrow$":

$$A(z) = \sum_{j=0}^{\infty} a_j z^j \text{ analytic on } \{|z| < 1\},$$
$$\Rightarrow A \in H^2$$

Hence Lemma 8.1 implies

$$\int_{-\pi}^{\pi} \log\left|A\left(e^{-i\lambda}\right)\right|^2 d\lambda > -\infty$$

2) "$\Leftarrow$"

$$\int_{-\pi}^{\pi} \log f_X(\lambda)\, d\lambda > -\infty$$

Lemma 8.2 implies

$$f(\lambda) = \frac{1}{2\pi} \left|\bar{g}\left(e^{-i\lambda}\right)\right|^2 \ (\lambda_{\text{Lebesgue}} - a.s.)$$

for some

$$g(z) = \sum_{j=0}^{\infty} g_j z^j \in H^2.$$

Thus, setting

$$A\left(e^{-i\lambda}\right) = \sum_{j=0}^{\infty} a_j z^j = \sum_{j=0}^{\infty} g_j z^j$$

we have

$$X_t = \sum_{j=0}^{\infty} a_j \varepsilon_{t-j}.$$

Remark 8.1 Let

$$S = \{\lambda \in [-\pi, \pi] \text{ s.t.} f(\lambda) = 0\}$$

Then

$$\int_{-\pi}^{\pi} \log f(\lambda)\, d\lambda > -\infty \Rightarrow \lambda_{\text{Lebesgue}}(S) = 0$$

but

$$\int_{-\pi}^{\pi} \log f(\lambda)\, d\lambda > -\infty \not\Leftarrow \lambda_{\text{Lebesgue}}(S) = 0$$

Remark 8.2

$$\int \log f(\lambda)\, d\lambda \leq \int f(\lambda)\, d\lambda < \infty.$$

Thus

$$\int \log f(\lambda)\, d\lambda > -\infty$$

implies

$$\left| \int \log f(\lambda)\, d\lambda \right| < \infty$$

Problem 8.4

$$\text{given} f \Rightarrow \text{Wold decomposition invertible?}$$

Solution 8.4 The solution can be derived using results from complex analysis.

Theorem 8.3 *Assumptions:*

$$X_t \ (t \in \mathbb{Z}) \text{ weakly stationary, spectral density } f_X$$

$$\int_{-\pi}^{\pi} \log f_X(\lambda)\, d\lambda > -\infty$$

Then

$$\exists! a_j \in \mathbb{C} \ (j = 0, 1, 2, \ldots) \text{ s.t.:}$$

$$\mathbb{R} \ni a_o > 0, \ \sum_{j=0}^{\infty} |a_j|^2 < \infty,$$

$$A(z) = \sum_{j=0}^{\infty} a_j z^j \neq 0 \ (|z| < 1),$$

$$f_X(\lambda) = \left|A\left(e^{-i\lambda}\right)\right|^2$$

Proof See e.g. Doob (1953).

Remark 8.3

$$\sum_{j=0}^{\infty} |a_j|^2 < \infty \Rightarrow A(z) \ (|z| < 1) \text{ analytic}$$

Remark 8.4 It is essential that the power series $\sum a_j z^j$ is one sided so that we get the one-sided Wold decomposition.

Theorem 8.4 *Assumptions:*

$$X_t \ (t \in \mathbb{Z}) \text{ weakly stationary, spectral density } f_X$$

$$\int_{-\pi}^{\pi} \log f_X(\lambda)\, d\lambda > -\infty$$

$$A(z) \text{ as in Theorem 8.3 analytic for } |z| \leq 1$$

$$A(z) \neq 0 \ (|z| = 1)$$

Then

$$X_t = A(B)\, \varepsilon_t = \text{invertible Wold decomposition}$$

Proof Theorem 8.3 implies

$$A(z) \neq 0 \ (|z| < 1).$$

The additional assumptions are

$$A(z) \neq 0\ (|z| = 1), A(z) \text{ analytic for } |z| \leq 1.$$

Hence

$$\exists \delta > 0 \text{ s.t. } A(z) \neq 0 \text{ analytic for } |z| < 1 + \delta.$$

Therefore,

$$\frac{1}{A(z)} \text{ analytic for } |z| < 1 + \delta,$$

$$\frac{1}{A(z)} = \sum_{j=0}^{\infty} b_j z^j, \sum_{j=0}^{\infty} |b_j| < \infty, \frac{1}{A(e^{-i\lambda})} \in L^2(F_X).$$

Then

$$A^{-1}(B)X_t = \int e^{it\lambda} \frac{1}{A(e^{-i\lambda})} dZ_X(\lambda; \omega)$$

is well defined,

$$A(B)[A^{-1}(B)X_t] = X_t$$

and

$$\varepsilon_t = A^{-1}(B)X_t = \sum_{j=0}^{\infty} b_j X_{t-j}$$

8.3 Construction of the Wold Decomposition from *f*

Problem 8.5

$$f \Rightarrow \text{calculate Wold decomposition}$$

Solution 8.5 Apply results from complex analysis.

Theorem 8.5 (Whittle 1962) *Assumptions:*

$$X_t \ (t \in \mathbb{Z}, \text{ weakly stationary, } E(X_t) = 0, \text{ spectral density } f_X$$

$$\int_{-\pi}^{\pi} \log f_X(\lambda)\, d\lambda > -\infty$$

$$G(z) = \sum_{k=-\infty}^{\infty} \gamma_X(k) z^k$$

$$\exists r > 1 \text{ s.t. } L(z) = \log G(z) \text{ holomorphic in } r^{-1} < |z| < r$$

Then X_t has the Wold representation

$$X_t = A(B)\varepsilon_t = \sum_{j=0}^{\infty} a_j \varepsilon_{t-j},\ a_0 = 1$$

with

$$A(z) = \sum_{j=0}^{\infty} a_j z^j = \exp\left(\sum_{j=1}^{\infty} \alpha_j z^j\right)$$

$$\alpha_j = \frac{1}{2\pi}\int_{-\pi}^{\pi} e^{ij\lambda} \log f_X(\lambda)\, d\lambda,$$

$$\varepsilon_t = \int e^{it\lambda} \frac{1}{A(e^{-i\lambda})} dZ_X(\lambda) = \sum_{j=0}^{\infty} b_j X_{t-j}.$$

Moreover,

$$\sum_{j=0}^{\infty} b_j z^j = \frac{1}{A(z)},$$

$$f(\lambda) = \exp(\alpha_0)\left|A\left(e^{-i\lambda}\right)\right|^2,$$

$$\sigma_\varepsilon^2 = var(\varepsilon_t) = 2\pi \exp(\alpha_0). \tag{8.4}$$

Proof

1)

$$L(z) = \log \frac{G(z)}{2\pi} \text{ holomorphic in } r^{-1} < |z| < r$$

implies

$$\text{Laurent series } L(z) = \sum_{j=-\infty}^{\infty} \alpha_j z^j,$$

$$\frac{G(z)}{2\pi} = \exp L(z) \text{ holomorphic in } r^{-1} < |z| < r,$$

and

$$r = \limsup_{n\to\infty} \frac{1}{\sqrt[n]{|\alpha_n|}} > 1 \Rightarrow \sum_{j=-\infty}^{\infty} |\alpha_j| < \infty.$$

Defining

$$A(z) = \exp\left(\sum_{j=1}^{\infty} \alpha_j z^j\right), \tilde{A}(z) = \exp\left(\sum_{j=-\infty}^{-1} \alpha_j z^j\right)$$

we have

$$\frac{G(z)}{2\pi} = \exp L(z) = \exp\left(\sum_{j=-\infty}^{\infty} \alpha_j z\right) = \exp(\alpha_0) A(z) \tilde{A}(z).$$

By definition

$$A(0) = \tilde{A}(0) = 1.$$

In particular, for $z = \exp(-i\lambda)$,

$$\frac{G\left(e^{-i\lambda}\right)}{2\pi} = f_X(\lambda) = \exp(\alpha_0) A\left(e^{-i\lambda}\right) \tilde{A}\left(e^{-i\lambda}\right)$$

and

$$\log f_X(\lambda) = \alpha_o + \sum_{j=1}^{\infty} \alpha_j e^{-ij\lambda} + \sum_{j=-\infty}^{-1} \alpha_j e^{-ij\lambda}.$$

Now

$$\sum_{j=-\infty}^{\infty} |\alpha_j| < \infty$$

implies

$$\frac{1}{2\pi}\int_{-\pi}^{\pi} e^{ij\lambda}\log f_X(\lambda)\,d\lambda = \frac{1}{2\pi}\sum_{l=-\infty}^{\infty}\alpha_l\int_{-\pi}^{\pi} e^{i(j-l)\lambda}d\lambda = \alpha_j.$$

Moreover,

$$\log f_X(\lambda) = \log f_X(-\lambda) \Rightarrow \alpha_j = \alpha_{-j} \in \mathbb{R} \Rightarrow \tilde{A}(z) = A\left(z^{-1}\right)$$

and

$$|z| = 1 \Rightarrow z^{-1} = \bar{z} \Rightarrow \tilde{A}(z) = \overline{A(z)}.$$

Hence

$$\begin{aligned} f_X(\lambda) &= \frac{G\left(e^{-i\lambda}\right)}{2\pi} \\ &= \exp(\alpha_0) A\left(e^{-i\lambda}\right)\overline{A(e^{-i\lambda})} \\ &= \exp(\alpha_0)\left|A\left(e^{-i\lambda}\right)\right|^2. \end{aligned}$$

so that

$$f(\lambda) = (2\pi)^{-1}G(e^{-i\lambda}) = e^{\alpha_o}|A(e^{-i\lambda})|^2.$$

2) By assumption,

$$f^*(z) = \frac{G(z)}{2\pi} \text{ holomorphic in } r^{-1} < |z| < r.$$

However, from 1) we have

$$f^*(z) = \exp(\alpha_0) A(z)\overline{A(z)},$$

$$A(z) = \sum_{j=0}^{\infty} a_j z^j, \overline{A(z)} = \sum_{j=0}^{\infty} \bar{a}_j \bar{z}^j, a_0 = 1$$

so that

$$f^*(z) = \exp(\alpha_0)\left(\sum_{j=0}^{\infty} a_j z^j\right)\left(\sum_{j=0}^{\infty} \bar{a}_j \bar{z}^j\right) \text{ holomorphic in } |z| < r,$$

$$\sum_{j=0}^{\infty}\left|a_j\right| < \infty,$$

Moreover,

$$A(z) = \exp\left(\sum_{j=1}^{\infty} \alpha_j z^j\right) \neq 0\,(|z| < r)$$

implies

$$\frac{1}{A(z)} = \exp\left(-\sum_{j=1}^{\infty} \alpha_j z^j\right) \text{ holomorphic in } |z| < r,$$

$$\frac{1}{A(z)} = \sum_{j=0}^{\infty} b_j z^j,\, b_0 = 1,$$

$$\sum_{j=0}^{\infty} \left|b_j\right| < \infty.$$

Therefore,

$$\varepsilon_t = \int e^{it\lambda} \frac{1}{A\left(e^{-i\lambda}\right)} dZ_X(\lambda) = \sum_{j=0}^{\infty} b_j X_{t-j} \in L^2(F_X),$$

$$cov(\varepsilon_s, \varepsilon_t) = 0\,(s \neq t),$$

$$X_t = A(B)\,\varepsilon_t = \sum_{j=0}^{\infty} a_j \varepsilon_{t-j}$$

3)

$$\begin{aligned}\sigma_\varepsilon^2 = var(\varepsilon_t) &= \int \frac{1}{\left|A\left(e^{-i\lambda}\right)\right|^2} f_X(\lambda)\, d\lambda \\ &= \int_{-\pi}^{\pi} \frac{1}{\left|A\left(e^{-i\lambda}\right)\right|^2} \exp(\alpha_0) \left|A\left(e^{-i\lambda}\right)\right|^2 d\lambda \\ &= 2\pi \exp(\alpha_o)\end{aligned}$$

Remark 8.5

$$f_X(\lambda) = \exp(\alpha_0) \left|A\left(e^{-i\lambda}\right)\right|^2$$

is called canonical factorization of f_X,

$$\sigma_\varepsilon^2 = 2\pi \exp(\alpha_o) = 2\pi \exp\left(\frac{1}{2\pi} \int_{-\pi}^{\pi} \log f_X(\lambda)\, d\lambda\right)$$

is called Wiener-Kolmogorov formula for the one-step prediction error (Wiener and Masani 1958; Kolmogorov 1941a,b). Note that

$$\int_{-\pi}^{\pi} \log f_X(\lambda)\, d\lambda = -\infty \Rightarrow \sigma_\varepsilon^2 = 0$$

8.4 Best Linear Prediction Given a Finite Past

Problem 8.6

$$X_t \text{ weakly stationary, purely stochastic, } E(X_t) = 0, \text{ acf } \gamma_X(k) \tag{8.5}$$

Find the optimal linear prediction $\hat{X}_{n+k}$ of X_{n+k} given X_s $(1 \le s \le n)$.

Solution 8.6 Orthogonal projection of X_{n+k} on L^2-closure of space spanned by $X_1, \ldots, X_n$.

Definition 8.2

$$L_{m,n}^0 = \left\{ Y \mid Y = \sum_{j=1}^{k} a_j X_{t_j},\ k \in \mathbb{N},\ a_j \in \mathbb{R},\ t_j \in \mathbb{Z},\ m \le t_j \le n \right\},$$

$$L_{m,n} = \overline{L_{m,n}^0} = \left\{ Y \mid \exists Y_i \in L_{m,n}^0\ (i \in \mathbb{N}) \text{ s.t. } \lim_{i\to\infty} \|Y - Y_i\|_{L^2(\Omega)}^2 = 0 \right\}.$$

Definition 8.3

$$P_{L_{m,n}}(X_{n+k}) = \text{orthogonal projection of } X_{n+k} \text{ on } L_{m,n}$$

Lemma 8.3 *Under (8.5),*

$$\exists \varphi_{n,j}(k) \in \mathbb{R}\ (j = 1, \ldots, n) \text{ s.t. } P_{L_{1,n}}(X_{n+k}) = \sum_{j=1}^{n} \varphi_{n,j}(k)\, X_{n-j+1}$$

Proof Follows from Corollary 2.12.

Corollary 8.2 *Under (8.5), let*

$$\hat{X}_{n+k} = \sum_{j=1}^{n} \varphi_{n,j}(k)\, X_{n-j+1}.$$

Then

$$MSE_n(k) = E\left[\left(X_{n+k} - \hat{X}_{n+k}\right)^2\right] = \inf_{Y \in L_{1,n}} E\left[(X_{n+k} - Y)^2\right]$$

Proof Follows from

$$\sum_{j=1}^{n} \varphi_{n,j}(k) X_{n-j+1} = P_{L_{1,n}}(X_{n+k}).$$

Notation 21

$$\begin{aligned} \Sigma_n &= [\gamma(i-j)]_{i,j=1,2,\ldots,n} \\ \mathbf{X}_n &= (X_n, X_{n-1}, \ldots, X_1)^T \\ \mathbf{g}_n(k) &= [\gamma(k), \gamma(k+1), \ldots, \gamma(k+n-1)]^T \\ \boldsymbol{\varphi}_n(k) &= [\varphi_{n1}(k), \varphi_{n2}(k), \ldots, \varphi_{nn}(k)]^T \end{aligned}$$

Theorem 8.6 *Under (8.5), we have*

$$\begin{aligned} \boldsymbol{\varphi}_n(k) &= \Sigma_n^{-1} \mathbf{g}_n(k), \\ \hat{X}_{n+k} &= \boldsymbol{\varphi}_n^T(k) \mathbf{X}_n, \\ MSE_n(k) &= \gamma(0) - \mathbf{g}_n^T(k) \Sigma_n^{-1} \mathbf{g}_n(k). \end{aligned}$$

Proof

1) By definition

$$\left\langle X_{n+k} - \sum_{j=1}^{n} \varphi_{n,j}(k) X_{n+1-j}, X_{n+1-s} \right\rangle = 0 \ (s = 1, 2, \ldots, n)$$

$$\Rightarrow \gamma(k+s-1) = \sum_{j=1}^{n} \varphi_{n,j}(k) \gamma(s-j) \ (s = 1, 2, \ldots, n),$$

$$\Rightarrow \mathbf{g}_n(k) = \Sigma_n \boldsymbol{\varphi}_n(k).$$

Now

$$X_t \text{ purely stochastic} \Rightarrow \exists \Sigma_n^{-1}$$

$$\Rightarrow \boldsymbol{\varphi}_n(k) = \Sigma_n^{-1} \mathbf{g}_n(k)$$

2)

$$\hat{X}_{n+k} = \boldsymbol{\varphi}_n^T(k)\,\mathbf{X}_n = \mathbf{g}_n^T(k)\,\Sigma_n^{-1}\mathbf{X}_n$$

Hence,

$$\begin{aligned} MSE_n(k) &= E\left[\left(X_{n+k} - \hat{X}_{n+k}\right)^2\right] \\ &= \gamma(0) - 2\mathbf{g}_n^T(k)\,\Sigma_n^{-1}E\left[X_{n+k}\mathbf{X}_n\right] + \mathbf{g}_n^T(k)\,\Sigma_n^{-1}\mathbf{g}_n(k) \\ &= \gamma(0) - \mathbf{g}_n^T(k)\,\Sigma_n^{-1}\mathbf{g}_n(k) \end{aligned}$$

Corollary 8.3 *Under (8.5)*

$$MSE_n(k) \le \gamma(0)\ \ (k \ge 1)$$

with

$$MSE_n(k) < \gamma(0) \Leftrightarrow \max_{j=0,\ldots,n-1} |\gamma(k+j)| > 0$$

Proof

$$MSE_n(k) = \gamma(0)$$
$$\Leftrightarrow$$
$$\mathbf{g}_n(k) = \mathbf{0}$$
$$\Leftrightarrow$$
$$\gamma(k) = \ldots = \gamma(k+n-1) = 0$$

Problem 8.7 Do we need the entire past $X_n, X_{n-1}, \ldots, X_1$ for the linear prediction of X_{n+1}?

Solution 8.7 Measure additional linear contribution of X_1 to prediction of X_{n+1} after $X_n, X_{n-1}, \ldots, X_2$ have been taken into account already.

Definition 8.4

$$\rho_{\text{partial}}(n) = corr\left(X_1 - P_{L_{2,n}}(X_1), X_{n+1} - P_{L_{2,n}}(X_{n+1})\right)\ \ (n \in \mathbb{N})$$

is called partial autocorrelation function (pacf).

Lemma 8.4 *Under (8.5),*

$$\rho_{partial}(n) = \varphi_{nn}(1)$$

Proof

$$X_{n+1} = \hat{X}_{n+1} + \left(X_{n+1} - \hat{X}_{n+1}\right)$$
$$= \varphi_{n,n}(1) X_1 + \sum_{j=1}^{n-1} \varphi_{n,j}(1) X_{n-j+1} + \left(X_{n+1} - \hat{X}_{n+1}\right)$$

implies

$$P_{L_{2,n}}(X_{n+1}) = \varphi_{n,n}(1) P_{L_{2,n}}(X_1) + \sum_{j=1}^{n-1} \varphi_{n,j}(1) X_{n-j+1},$$

$$X_{n+1} - P_{L_{2,n}}(X_{n+1}) = \varphi_{n,n}(1)\left[X_1 - P_{L_{2,n}}(X_1)\right] + \left(X_{n+1} - \hat{X}_{n+1}\right)$$

and therefore

$$\left\langle X_1 - P_{L_{2,n}}(X_1), X_{n+1} - P_{L_{2,n}}(X_{n+1})\right\rangle = \varphi_{n,n}(1) \left\| X_1 - P_{L_{2,n}}(X_1)\right\|^2$$
$$= \varphi_{n,n}(1) \left\| X_{n+1} - P_{L_{2,n}}(X_{n+1})\right\|^2.$$

Hence,

$$\rho_{\text{partial}}(n) = corr\left(X_1 - P_{L_{2,n}}(X_1), X_{n+1} - P_{L_{2,n}}(X_{n+1})\right)$$
$$= \frac{\varphi_{n,n}(1)\left\|X_{n+1} - P_{L_{2,n}}(X_{n+1})\right\|^2}{\left\|X_1 - P_{L_{2,n}}(X_1)\right\| \left\|X_{n+1} - P_{L_{2,n}}(X_{n+1})\right\|}$$
$$= \varphi_{n,n}(1) \frac{\left\|X_{n+1} - P_{L_{2,n}}(X_{n+1})\right\|^2}{\left\|X_{n+1} - P_{L_{2,n}}(X_{n+1})\right\|^2}$$
$$= \varphi_{n,n}(1).$$

Corollary 8.4 *Assumption:*

$$X_t \text{ stationary } AR(p)\text{-process}, \ \varphi_p \neq 0$$

Then

$$\rho_{partial}(p) = \varphi_p, \ \rho_{partial}(n) = 0 \ (n \geq p+1). \tag{8.6}$$

Remark 8.6 Equation (8.6) can be used for identifying AR(p) processes using estimated partial autocorrelations $\hat{\rho}_{\text{partial}}(k)$ ($k = 1, 2, \ldots$).

Recursive calculation of $\varphi_{n,j}(1)$ and $MSE_n(1)$ (Durbin-Levinson algorithm):

Theorem 8.7 *Under (8.5),*

$$\varphi_{11}(1) = \rho(1), MSE_o(1) = \gamma(0)$$

and, for $n \geq 2$,

$$\varphi_{nn}(1) = \frac{\gamma(n) - \sum_{j=1}^{n-1} \varphi_{n-1,j}\gamma(n-j)}{MSE_{n-1}(1)},$$

$$\begin{pmatrix} \varphi_{n,1}(1) \\ \vdots \\ \varphi_{n,n-1}(1) \end{pmatrix} = \begin{pmatrix} \varphi_{n-1,1}(1) \\ \vdots \\ \varphi_{n-1,n-1}(1) \end{pmatrix} - \varphi_{n,n}(1) \begin{pmatrix} \varphi_{n-1,n-1}(1) \\ \vdots \\ \varphi_{n-1,1}(1) \end{pmatrix},$$

$$MSE_n(1) = MSE_{n-1}(1)\left[1 - \varphi_{n,n}^2(1)\right].$$

Proof Similarly arguments as in Lemma 8.4 (see e.g. Brockwell and Davis 1991).

Chapter 9
Inference for μ, γ and F

9.1 Location Estimation

Theorem 9.1 *Assumptions:*

$$X_t \in \mathbb{R} \ (t \in \mathbb{Z}) \text{ weakly stationary, } \mu = E(X_t)$$

$$f_X(\lambda) = \frac{1}{2\pi} \sum_{t=-\infty}^{\infty} e^{-ik\lambda} \gamma_X(k)$$

$$0 < f_X(0) < \infty$$

$$\bar{x} = n^{-1} \sum_{t=1}^{n} X_t$$

Then

$$var(\bar{x}) \underset{n\to\infty}{\sim} 2\pi f_X(0)\, n^{-1}.$$

Proof

$$var(\bar{x}) = n^{-1} \sum_{k=-(n-1)}^{n-1} \left(1 - \frac{|k|}{n}\right) \gamma_X(k) \underset{n\to\infty}{\sim} n^{-1} \sum_{k=-n}^{n} \gamma_X(k) \underset{n\to\infty}{\sim} n^{-1} 2\pi f_X(0).$$

J. Beran, *Mathematical Foundations of Time Series Analysis*,
https://doi.org/10.1007/978-3-319-74380-6_9

Theorem 9.2 *Assumptions:*

$$X_t \in \mathbb{R}\ (t \in \mathbb{Z})\ \text{weakly stationary},\ \mu = E(X_t)$$

$$f_X(\lambda) = \frac{1}{2\pi} \sum_{t=-\infty}^{\infty} e^{-ik\lambda} \gamma_X(k)$$

$$f_X(\lambda) \underset{\lambda \to 0}{\sim} c_f |\lambda|^{-d}\ (0 < c_f < \infty,\ -\frac{1}{2} < d < \frac{1}{2})$$

$$\nu(d) = \frac{2\Gamma(1-2d)\sin \pi d}{d(2d+1)}\ (d \neq 0)$$

$$\nu(0) = 2\pi$$

Then

$$var(\bar{x}) \underset{n \to \infty}{\sim} \nu(d) f_X\left(n^{-1}\right) n^{-1}.$$

Proof See e.g. Beran et al. (2013, Corollary 1.2).

Problem 9.1 Are there more efficient estimators than $\bar{x}$?

Solution 9.1 Compare $var(\bar{x})$ with the variance of the BLUE.

Notation 22

$$\mathbf{1} = \underset{n\times 1}{\mathbf{1}} = \left(1 \cdots 1\right)^T$$

$$\mathbf{w} = \left(w_1 \cdots w_n\right)^T$$

$$\mathbf{X}_n = \left(X_1 \cdots X_n\right)^T$$

$$\Sigma_n = \left[\gamma_X(i-j)\right]_{i,j=1,\ldots,n}$$

$$MSE(\hat{\mu}) = E\left[(\hat{\mu} - \mu)^2\right]$$

Definition 9.1 Let

$$X_t \in \mathbb{R}\ (t \in \mathbb{Z})\ \text{weakly stationary},\ \mu = E(X_t)$$

$$W_n = \left\{\mathbf{w} \mid \mathbf{w} \in \mathbb{R}^n, \mathbf{w}^T\mathbf{1} = 1\right\}.$$

Then

$$M_{n,\text{linear}} = \left\{\hat{\mu} \mid \hat{\mu} = \hat{\mu}_{\mathbf{w}} = \mathbf{w}^T\mathbf{X}_n,\, \mathbf{w} \in W_n\right\}$$

is the class of linear unbiased estimators of μ. If $\hat{\mu} \in M_{n,\text{linear}}$ and

$$MSE\,(\hat{\mu}) = \inf_{\hat{\mu}^* \in M_{n,\text{linear}}} MSE\left(\hat{\mu}^*\right),$$

then

$$\hat{\mu} = \hat{\mu}_{BLUE}$$

is the best linear unbiased estimator (BLUE) of μ.

Theorem 9.3 *Assumptions:*

$$X_t \in \mathbb{R}\ (t \in \mathbb{Z})\ \text{weakly stationary},\ \mu = E\,(X_t)$$

$$\mathbf{w} = \frac{\Sigma_n^{-1}\mathbf{1}}{\mathbf{1}^T\Sigma_n^{-1}\mathbf{1}}$$

Then

$$\hat{\mu} = \mathbf{w}^T\mathbf{X}_n = \hat{\mu}_{BLUE}$$

and

$$var\,(\hat{\mu}_{BLUE}) = \left(\mathbf{1}^T\Sigma_n^{-1}\mathbf{1}\right)^{-1}.$$

Proof Follows by defining the Lagrange function

$$L = \mathbf{w}^T\Sigma_n\mathbf{w} - \lambda\left(\mathbf{w}^T\mathbf{1} - 1\right)$$

and solving

$$\frac{\partial}{\partial\lambda}L = 0,\ \frac{\partial}{\partial\mathbf{w}}L = \mathbf{0}.$$

The question of relative efficiency of $\bar{x}$ compared to $\hat{\mu}_{BLUE}$ is answered more generally in the context of linear regression in the following section.

9.2 Linear Regression

Notation 23

$$u_j : \mathbb{N} \to \mathbb{R}\ (j = 0, 1, \ldots, p)$$

$$\mathbf{u}_j = \mathbf{u}_{j,n} = \left(u_j(1) \cdots u_j(n)\right)^T \in \mathbb{R}^n\ (j = 0, \ldots, p)$$

$$\underset{n\times(p+1)}{U} = \left(\mathbf{u}_0, \ldots, \mathbf{u}_p\right) \in M(n, p+1; \mathbb{R})$$

$$\mathbf{X}_n = \begin{pmatrix} X_1 \\ \vdots \\ X_n \end{pmatrix}, \mathbf{Y}_n = \begin{pmatrix} Y_1 \\ \vdots \\ Y_n \end{pmatrix}, \beta = \begin{pmatrix} \beta_0 \\ \vdots \\ \beta_p \end{pmatrix}$$

$$\left\|\mathbf{u}_{j,n}\right\|^2 = \sum_{t=1}^{n} u_j^2(t),$$

$$\underset{(p+1)\times(p+1)}{D_n} = diag\left(\left\|\mathbf{u}_{0,n}\right\|, \ldots, \left\|\mathbf{u}_{p,n}\right\|\right)$$

Definition 9.2 Let

$$X_t\ (t \in \mathbb{Z}) \text{ weakly stationary, } E(X_t) = 0, \text{ acf } \gamma_X,$$

$$Y_t = \sum_{j=0}^{p} \beta_j u_j(t) + X_t\ (t = 1, \ldots, n).$$

Then the least squares estimator (LSE) of β is defined by

$$\hat{\beta}_{LSE} = \arg\min_{\beta^*} \left\|\mathbf{Y}_n - U_n\beta^*\right\|^2.$$

Moreover, the best linear estimator (BLUE) of β is defined by

$$\hat{\beta}_{BLUE} = \arg\min_{\beta^*} \left(\mathbf{Y}_n - U_n\beta^*\right)^T \Sigma_n^{-1} \left(\mathbf{Y}_n - U_n\beta^*\right).$$

Lemma 9.1

$$\hat{\beta}_{LSE} = \left(U_n^T U_n\right)^{-1} U_n^T \mathbf{Y}_n,$$

$$E\left(\hat{\beta}_{LSE}\right) = \beta,$$

$$V_{n,LSE} = var\left(\hat{\beta}_{LSE}\right) = \left(U_n^T U_n\right)^{-1} U_n^T \Sigma_n U_n \left(U_n^T U_n\right)^{-1}.$$

Proof Follows from solving

$$\frac{\partial}{\partial \beta}\left[\left(\mathbf{Y}_n - U_n\hat{\beta}\right)^T\left(\mathbf{Y}_n - U_n\hat{\beta}\right)\right] = 0$$

Lemma 9.2

$$\hat{\beta}_{BLUE} = \left(U_n^T \Sigma^{-1} U_n\right)^{-1} U_n^T \Sigma_n^{-1} \mathbf{Y}_n,$$

$$E\left(\hat{\beta}_{BLUE}\right) = \beta,$$

$$V_{n,BLUE} = var\left(\hat{\beta}_{BLUE}\right) = \left(U_n^T \Sigma_n^{-1} U_n\right)^{-1}.$$

Proof Follows from solving

$$\frac{\partial}{\partial \beta}\left[\left(\mathbf{Y}_n - U_n\hat{\beta}\right)^T \Sigma_n^{-1}\left(\mathbf{Y}_n - U_n\hat{\beta}\right)\right] = 0$$

Problem 9.2 How much more efficient is the BLUE?

Solution 9.2 Grenander and Rosenblatt (1957) obtained an asymptotic solution under the assumption $0 < f_X(0) < \infty$ and additional regularity conditions.

Conditions (Grenander and Rosenblatt 1957):

(R1)

$$\forall j \in \{0, 1, \ldots, p\} : \lim_{n\to\infty} \left\|\mathbf{u}_{j,n}\right\|^2 = \infty$$

(R2)

$$\forall j \in \{0, 1, \ldots, p\} : \lim_{n\to\infty} \frac{\left\|\mathbf{u}_{j,n}\right\|^2}{\left\|\mathbf{u}_{j,n-1}\right\|^2} = 1.$$

(R3)

$$\forall j, l \in \{0, 1, \ldots, p\}, k \in \mathbb{N} : \exists \rho_{jl}(k) = \lim_{n\to\infty} \frac{\sum_{t=1}^{n} u_j(t)\, u_l(t+k)}{\left\|\mathbf{u}_{j,n}\right\| \left\|\mathbf{u}_{l,n}\right\|} \in \mathbb{R}.$$

(R4)

$$R(k) := \left[\rho_{jl}(k)\right]_{j,l=0,\ldots,p} \quad (k \in \mathbb{Z}),$$

$$\rho_{jl}(-k) = \rho_{lj}(k) \quad (k \in \mathbb{N}).$$

Then $R(0)$ is non-singular.

Remark 9.1

$$R(0) = R^T(0)$$

Lemma 9.3

$$\exists G : [-\pi, \pi] \to M(p+1, p+1, \mathbb{C}) \text{ s.t.:}$$

$$\forall \lambda \in [-\pi, \pi] : G(\lambda) = \overline{G^T(\lambda)} \text{ (Hermitian)},$$

$$\forall \lambda_1 \leq \lambda_2 \in [-\pi, \pi] : G(\lambda_2) - G(\lambda_1) \geq 0 \text{ (positive semidefinite increments)}$$

and

$$R(k) = \int_{-\pi}^{\pi} e^{ik\lambda} dG(\lambda) \quad (k \in \mathbb{Z}) \tag{9.1}$$

Proof Similar to the spectral representation of $\gamma_X(k)$. See Grenander and Rosenblatt (1957) for a detailed proof.

Definition 9.3 The function G in (9.1) is called spectral distribution function of the regression functions $u_0, \ldots, u_p$.

Lemma 9.4 *Assumptions: (R1), (R2), (R3), (R4),*

$$X_t \ (t \in \mathbb{Z}) \text{ weakly stationary}$$

$$E(X_t) = 0, \text{ acf } \gamma_X, \text{ spectral density } f_X$$

$$\sum_{k=-\infty}^{\infty} |\gamma_X(k)| < \infty \tag{9.2}$$

$$Y_t = \sum_{j=0}^{p} \beta_j u_j(t) + X_t \ (t = 1, 2, \ldots, n)$$

Then

$$\exists V_{LSE} = \lim_{n\to\infty} D_n V_{n,LSE} D_n$$

and

$$V_{LSE} = G^{-1}(\pi) \left[2\pi \int_{-\pi}^{\pi} f_X(\lambda) dG(\lambda) \right] G^{-1}(\pi).$$

Proof For a detailed proof see Grenander and Rosenblatt (1957). Note that

$$V_{n,LSE} = \left(U_n^T U_n\right)^{-1} W_n \left(U_n^T U_n\right)^{-1}$$

where

$$W_n = \left[w_{jl}\right]_{j,l=0,\ldots,p} = U_n^T \Sigma_n U_n.$$

From (R3), and Lemma 9.3 we obtain

$$\lim_{n\to\infty} \frac{\left(U_n^T U_n\right)_{jl}}{\left\|\mathbf{u}_{j,n}\right\| \left\|\mathbf{u}_{l,n}\right\|} = \lim_{n\to\infty} \frac{\sum_{t=1}^{n} u_j(t)\, u_l(t)}{\left\|\mathbf{u}_{j,n}\right\| \left\|\mathbf{u}_{l,n}\right\|} = \rho_{jl}(0)$$

and hence

$$\lim_{n\to\infty} D_n^{-1} \left(U_n^T U_n\right) D_n^{-1} = R(0) = G(\pi).$$

Moreover,

$$\begin{aligned} w_{jl} &= \sum_{t,s=1}^{n} u_j(t)\, \gamma_X(t-s)\, u_l(s) \\ &= \sum_{k=0}^{n-1} \gamma_X(k) \sum_{t=1}^{n-k} u_j(t)\, u_l(t+k) + \sum_{k=-(n-1)}^{-1} \gamma_X(k) \sum_{t=-k+1}^{n} u_j(t)\, u_l(t+k). \end{aligned}$$

Assumptions (R1), (R2) and (9.2) then imply

$$\begin{aligned} \lim_{n\to\infty} \frac{w_{jl}}{\left\|\mathbf{u}_{j,n}\right\| \left\|\mathbf{u}_{l,n}\right\|} &= \lim_{n\to\infty} \sum_{k=-(n-1)}^{n-1} \gamma_X(k) \left(\frac{\sum_{t=1}^{n} u_j(t)\, u_l(t+k)}{\left\|\mathbf{u}_{j,n}\right\| \left\|\mathbf{u}_{l,n}\right\|} \right) \\ &= \sum_{k=-\infty}^{\infty} \gamma_X(k)\, \rho_{jl}(k) \\ &= \int_{-\pi}^{\pi} \left(\sum_{k=-\infty}^{\infty} \gamma_X(k)\, e^{ik\lambda} \right) dG_{jl}(\lambda) \\ &= 2\pi \int_{-\pi}^{\pi} f_X(\lambda)\, dG_{jl}(\lambda) \end{aligned}$$

and hence

$$\lim_{n\to\infty} D_n^{-1} W_n D_n^{-1} = 2\pi \int_{-\pi}^{\pi} f_X(\lambda)\, dG(\lambda).$$

Thus

$$\lim_{n\to\infty} D_n V_{n,LSE} D_n = \lim_{n\to\infty} \left[D_n \left(U_n^T U_n\right)^{-1} D_n\right] D_n^{-1} W_n D_n^{-1} \left[D_n \left(U_n^T U_n\right)^{-1} D_n\right]$$
$$= G^{-1}(\pi) \left[2\pi \int_{-\pi}^{\pi} f_X(\lambda)\, dG(\lambda)\right] G^{-1}(\pi).$$

Lemma 9.5 *Assumptions: (R1), (R2), (R3), (R4),*

$$X_t\ (t \in \mathbb{Z})\ \textit{weakly stationary}$$

$$E(X_t) = 0,\ \textit{acf}\ \gamma_X,\ \textit{spectral density}\ f_X$$

$$\sum_{k=-\infty}^{\infty} |\gamma_X(k)| < \infty \tag{9.3}$$

$$f_X\ \textit{piecewise continuous},\ f_X > 0 \tag{9.4}$$

$$Y_t = \sum_{j=0}^{p} \beta_j u_j(t) + X_t\ (t = 1, 2, \dots, n)$$

Then

$$\exists V_{BLUE} = \lim_{n\to\infty} D_n V_{n,BLUE} D_n$$

and

$$V_{BLUE} = \lim_{n\to\infty} D_n V_{n,BLUE} D_n = \left(\frac{1}{2\pi} \int_{-\pi}^{\pi} \frac{1}{f_X(\lambda)} dG(\lambda)\right)^{-1}.$$

Proof For a detailed proof see Grenander and Rosenblatt (1957). Note that

$$V_{n,BLUE} = W_n^{-1}$$

where

$$W_n = \left[w_{jl}\right]_{j,l=0,\dots,p} = U_n^T \Sigma_n^{-1} U_n.$$

Let

$$\gamma_X^*(k) = (2\pi)^{-2} \int_{-\pi}^{\pi} e^{ik\lambda} \frac{1}{f_X(\lambda)} d\lambda,$$

$$\Sigma_n^* = \left[\gamma_X^*(j-l)\right]_{j,l=1,\ldots,n},$$

$$W_n^* = \left[w_{jl}^*\right]_{j,l=0,\ldots,p} = U_n^T \Sigma_n^* U_n.$$

Then (9.3) and (9.4) imply

$$\lim_{n\to\infty} D_n^{-1} W_n D_n^{-1} = \lim_{n\to\infty} D_n^{-1} W_n^* D_n^{-1}.$$

By analogous arguments as in the proof of Lemma 9.4,

$$\begin{aligned}
\lim_{n\to\infty} \left[D_n^{-1} W_n^* D_n^{-1}\right]_{jl} &= \lim_{n\to\infty} \frac{\sum_{t,s=1}^{n} u_{j,n}(t)\, \gamma_X^*(t-s)\, u_{l,n}(s)}{\left\|\mathbf{u}_{j,n}\right\| \left\|\mathbf{u}_{l,n}\right\|} \\
&= \lim_{n\to\infty} \sum_{k=-(n-1)}^{n-1} \gamma_X^*(k) \left(\frac{\sum_{t=1}^{n} u_j(t)\, u_l(t+k)}{\left\|\mathbf{u}_{j,n}\right\| \left\|\mathbf{u}_{l,n}\right\|}\right) \\
&= \sum_{k=-\infty}^{\infty} \gamma_X^*(k)\, \rho_{jl}(k) \\
&= \int_{-\pi}^{\pi} \left(\sum_{k=-\infty}^{\infty} e^{ik\lambda} \gamma_X^*(k)\right) dG_{jl}(\lambda) \\
&= \frac{1}{2\pi} \int_{-\pi}^{\pi} \frac{1}{f_X(\lambda)} dG_{jl}(\lambda).
\end{aligned}$$

Hence,

$$\lim_{n\to\infty} D_n V_{n,BLUE} D_n = \lim_{n\to\infty} \left(D_n^{-1} W_n^* D_n^{-1}\right)^{-1} = \left(\frac{1}{2\pi} \int_{-\pi}^{\pi} \frac{1}{f_X(\lambda)} dG(\lambda)\right)^{-1}.$$

Notation 24

$$\Delta G(\lambda) = \lim_{\delta \downarrow 0} \left[G(\lambda+\delta) - G(\lambda-\delta)\right]$$

$$G(A) = \int_{\lambda \in A} dG(\lambda)$$

Definition 9.4

$$S = \{\lambda \in [0, \pi] \text{ s.t. } \Delta G(\lambda) > 0\}$$
$$= \text{regression spectrum}$$

Theorem 9.4 *Assume (R1) to (R4). Then*

$$\exists m \leq p \text{ and } S_0, \ldots, S_m \subseteq [-\pi, \pi] \text{ s.t.:}$$

$$S = \cup_{j=0}^{m} S_j,\ S_j \cap S_l = \phi\ (j \neq l),$$

$$G(S_j)\, G^{-1}(\pi)\, G(S_j) = G(S_j),$$

$$G(S_j)\, G^{-1}(\pi)\, G(S_l) = 0\ (j \neq l).$$

Proof See Grenander and Rosenblatt (1957).

Definition 9.5

$$S_0, \ldots, S_m = \text{elements of the regression spectrum}$$

Corollary 9.1 *Under the assumptions of Lemma 9.5, we have*

$$V_{BLUE} V_{LSE}^{-1} = 1$$

$$\Leftrightarrow$$

$$\exists c_j \in [0, \infty) \text{ s.t. } f_X(\lambda) \equiv c_j\ (\lambda \in S_j)$$

Proof

1) "$\Leftarrow$":

$$V_{BLUE} = \left(\frac{1}{2\pi} \int_{-\pi}^{\pi} \frac{1}{f_X(\lambda)} dG(\lambda)\right)^{-1},$$

$$V_{LSE} = G^{-1}(\pi) \left[2\pi \int_{-\pi}^{\pi} f_X(\lambda)\, dG(\lambda)\right] G^{-1}(\pi).$$

Then

$$f_X(\lambda) \equiv c_j\ (\lambda \in S_j, j = 0, \ldots, p)$$

implies

$$\int_{-\pi}^{\pi} f_X(\lambda)\, dG(\lambda) = \sum_{j=0}^{m} c_j G(S_j),$$

$$\int_{-\pi}^{\pi} \frac{1}{f_X(\lambda)} dG(\lambda) = \sum_{j=0}^{m} c_j^{-1} G(S_j).$$

Hence,

$$\begin{aligned} V_{LSE} V_{BLUE}^{-1} &= G^{-1}(\pi) \left(\sum_{j=0}^{m} c_j G(S_j) \right) G^{-1}(\pi) \left(\sum_{j=0}^{m} c_j^{-1} G(S_j) \right) \\ &= G^{-1}(\pi) \sum_{j,l=0}^{m} c_j c_l^{-1} G(S_j) G^{-1}(\pi) G(S_l) \\ &= G^{-1}(\pi) \sum_{j=0}^{m} G(S_j) = G^{-1}(\pi) G(\pi) = I. \end{aligned}$$

2) "$\Rightarrow$": see Grenander and Rosenblatt (1957).

Remark 9.2 Results can be generalized to $u_j(t) \in \mathbb{C}$, with $u_j u_l$ replaced by $u_j \bar{u}_l$.

Example 9.1 Let

$$Y_t = \beta_o + \beta_1 t + \ldots + \beta_p t^p + X_t.$$

Then

$$\left\| \mathbf{u}_{j,n} \right\|^2 = \sum_{t=1}^{n} t^{2j} = n^{2j+1} \sum_{t=1}^{n} \left(\frac{t}{n} \right)^{2j} \frac{1}{n} \sim n^{2j+1} \int_o^1 x^{2j} dx = \frac{n^{2j+1}}{2j+1},$$

$$\begin{aligned} \rho_{jl}(k) &= \lim_{n\to\infty} n^{-j-l-1} \sqrt{(2j+1)(2l+1)} \sum_{t=1}^{n} t^j (t+k)^l \\ &= \sqrt{(2j+1)(2l+1)} \lim_{n\to\infty} \sum_{t=1}^{n} \left(\frac{t}{n} \right)^j \left(\frac{t+k}{n} \right)^l \frac{1}{n} \\ &= \sqrt{(2j+1)(2l+1)} \int_0^1 x^{j+l} dx \\ &= \frac{\sqrt{(2j+1)(2l+1)}}{j+l+1}. \end{aligned}$$

Hence, $\rho_{jl}(k)$ does not depend on k, so that

$$\rho_{jl}(k) = \frac{\sqrt{(2j+1)(2l+1)}}{j+l+1} = \int_{-\pi}^{\pi} e^{ik\lambda} dG_{jl}(\lambda) \;\; (k \in \mathbb{Z})$$

implies

$$S = \{0\}.$$

Since (9.4) implies

$$0 < f_X(0) < \infty,$$

we obtain

$$V_{BLUE} = V_{LSE}.$$

Example 9.2 Let

$$\lambda_0, \ldots, \lambda_p \in [-\pi, \pi], \lambda_j \neq \lambda_l \; (j \neq l),$$

$$Y_t = \beta_0 e^{it\lambda_0} + \ldots + \beta_p e^{it\lambda_p} + X_t.$$

Then

$$\left\|\mathbf{u}_{j,n}\right\|^2 = n \; (j = 0, \ldots, n),$$

$$\rho_{jj}(k) = e^{-ik\lambda_j} = \overline{\rho_{jj}(-k)},$$

$$\rho_{jl}(k) = e^{-ik\lambda_l} \lim_{n\to\infty} n^{-1} \sum_{t=1}^{n} e^{it(\lambda_j - \lambda_l)} = 0 \; (j \neq l).$$

Then

$$\int_{-\pi}^{\pi} e^{ik\lambda} dG_{jj}(\lambda) = e^{-ik\lambda_j} \; (k \in \mathbb{Z})$$

and

$$\int_{-\pi}^{\pi} e^{ik\lambda} dG_{jl}(\lambda) = 0 \; (k \in \mathbb{Z}, j \neq l)$$

imply

$$S = \cup_{j=0}^{p} S_j \text{ with } S_j = \{\lambda_j\} \;\; (j = 0, \ldots, p).$$

Since (9.4) implies

$$0 < f_X(\lambda_j) < \infty \; (j = 0, \ldots, p),$$

we obtain

$$V_{BLUE} = V_{LSE}.$$

Corollary 9.2 *Assumptions:*

$$X_t \in \mathbb{R} \; (t \in \mathbb{Z}) \text{ weakly stationary, } \mu = E(X_t)$$

$$f_X(\lambda) = \frac{1}{2\pi} \sum_{t=-\infty}^{\infty} e^{-ik\lambda} \gamma_X(k)$$

$$0 < f_X(0) < \infty$$

Then

$$\lim_{n\to\infty} \frac{var(\hat{\mu}_{BLUE})}{var(\bar{x})} = 1.$$

Remark 9.3 The results above are not applicable to the case of long memory (see e.g. Adenstedt 1974; Beran and Künsch 1985; Samarov and Taqqu 1988; Yajima 1988, 1991; Dahlhaus 1995)

Remark 9.4 An alternative to parametric trend modelling is a nonparametric trend, i.e.

$$Y_t = \mu(t/n) + X_t,$$

with X_t stationary, $E(X_t) = 0$ and μ for instance twice continuously differentiable. For kernel and local polynomial estimation of μ in the time series context see e.g. Ghosh (2017) and references therein.

9.3 Nonparametric Estimation of γ

Definition 9.6

$$\hat{\gamma} : \mathbb{Z} \to \mathbb{R}, k \to \hat{\gamma}(k)$$

where

$$\hat{\gamma}(k) = 0 \ (|k| \geq n),$$

$$\hat{\gamma}(k) = \frac{1}{n}\sum_{t=1}^{n-k}(X_t - \bar{x})(X_{t+k} - \bar{x}) \ \ (0 \leq k \leq n-1),$$

$$\hat{\gamma}(k) = \hat{\gamma}(-k)$$

is called empirical autocovariance function. Moreover,

$$\hat{\rho}(k) = \frac{\hat{\gamma}(k)}{\hat{\gamma}(0)} \ (k \in \mathbb{Z}).$$

is called empirical autocorrelation function.

Problem 9.3 Is $\hat{\gamma}$ an autocovariance function?

Solution 9.3 Check positive semidefiniteness using the periodogram.

Definition 9.7

$$I(\lambda) = \frac{1}{2\pi n}\left|\sum_{t=1}^{n} X_t e^{it\lambda}\right|^2 = \text{periodogram}$$

$$I^0(\lambda) = \frac{1}{2\pi n}\left|\sum_{t=1}^{n}(X_t - \bar{x})\, e^{it\lambda}\right|^2 = \text{mean adjusted periodogram}$$

Definition 9.8

$$\lambda_j = \frac{2\pi j}{n} \ (j = 1, 2, \ldots, \left[\frac{n-1}{2}\right])$$

are called Fourier frequencies.

Lemma 9.6 *For Fourier frequencies λ_j we have*

$$I(\lambda_j) = I^0(\lambda_j)$$

Proof

$$\begin{aligned}\sum_{t=1}^{n} \bar{x} \exp\left(it\lambda_j\right) &= \bar{x} \sum_{t=1}^{n} \exp\left(it\lambda_j\right) \\ &= \bar{x} \exp\left(i\lambda_j\right) \frac{1 - \exp\left(i2\pi j\right)}{1 - \exp\left(i\lambda_j\right)} \\ &= 0\end{aligned}$$

Lemma 9.7

$$I(0) = \frac{n}{2\pi} \bar{x}^2, \; I^0(0) = 0$$

Proof

$$I(0) = \frac{1}{2\pi n} \left| \sum_{t=1}^{n} X_t \right|^2 = \frac{n}{2\pi} \left(\frac{1}{n} \sum_{t=1}^{n} X_t \right)^2$$

$$I^0(0) = \frac{1}{2\pi n} \left| \sum_{t=1}^{n} (X_t - \bar{x}) \right|^2 = 0$$

Lemma 9.8

$$\hat{\gamma}(k) = \int_{-\pi}^{\pi} e^{ik\lambda} I^0(\lambda)\, d\lambda \; (k \in \mathbb{Z})$$

and

$$I^0(\lambda) = \frac{1}{2\pi} \sum_{k=-(n-1)}^{n-1} \hat{\gamma}(k)\, e^{-ik\lambda} \; (\lambda \in [-\pi, \pi]).$$

Proof

1)

$$I^0(\lambda) = \frac{1}{2\pi n} \sum_{t,s=1}^{n} (X_t - \bar{x})(X_s - \bar{x})\, e^{i(t-s)\lambda}$$

implies, for $0 \le k \le n-1$,

$$\int_{-\pi}^{\pi} e^{ik\lambda} I^0(\lambda)\, d\lambda = \frac{1}{2\pi n} \sum_{t,s=1}^{n} (X_s - \bar{x})(X_t - \bar{x}) \int_{-\pi}^{\pi} e^{i(t-s+k)\lambda} d\lambda$$

$$= \frac{1}{n} \sum_{t=1}^{n-k} (X_t - \bar{x})(X_{t+k} - \bar{x})$$

$$= \hat{\gamma}(k),$$

and, for $|k| \ge n$,

$$\int_{-\pi}^{\pi} e^{ik\lambda} I^0(\lambda)\, d\lambda = 0 = \hat{\gamma}(k).$$

2)

$$I^0(\lambda) = \frac{1}{2\pi n} \sum_{t,s=1}^{n} (X_t - \bar{x})(X_s - \bar{x}) e^{i(t-s)\lambda}$$

$$= \frac{1}{2\pi n} \left[\sum_{k=0}^{n-1} e^{-ik\lambda} \sum_{t=1}^{n-k} (X_t - \bar{x})(X_{t+k} - \bar{x}) \right.$$

$$\left. + \sum_{k=-(n-1)}^{-1} e^{-ik\lambda} \sum_{t=1-k}^{n} (X_t - \bar{x})(X_{t+k} - \bar{x}) \right]$$

$$= \frac{1}{2\pi} \sum_{k=-(n-1)}^{n-1} e^{-ik\lambda} \frac{1}{n} \sum_{t=1}^{n-k} (X_t - \bar{x}) \left(X_{t+|k|} - \bar{x}\right)$$

$$= \frac{1}{2\pi} \sum_{k=-(n-1)}^{n-1} e^{-ik\lambda} \hat{\gamma}(k).$$

Corollary 9.3

$$\hat{\gamma} : \mathbb{Z} \to \mathbb{R}$$

as defined in Definition 9.6 is an autocovariance function.

Proof Corollary 3.10 and

$$I^0(\lambda) \ge 0,\ I^0(-\lambda) = I^0(\lambda)$$

imply

$$\hat{\gamma}(k)=\int_{-\pi}^{\pi}e^{ik\lambda}I^{0}(\lambda)\,d\lambda=\text{autocovariance function.}$$

Lemma 9.9 *For Fourier frequencies λ_j we have*

$$\hat{\gamma}(k)=\frac{2\pi}{n}\sum_{j=1}^{n}e^{ik\lambda_j}I\left(\lambda_j\right)=\frac{2\pi}{n}\sum_{j=-m}^{m}e^{ik\lambda_j}I^{0}\left(\lambda_j\right)\ (m=[(n-1)/2])$$

Proof Let $k\geq 0$. Then

$$I\left(\lambda_j\right)=I^{0}\left(\lambda_j\right)=\frac{1}{2\pi n}\sum_{t,s=1}^{n}\left(X_t-\bar{x}\right)\left(X_s-\bar{x}\right)e^{i(t-s)\lambda_j}$$

and

$$\sum_{j=1}^{n}e^{i(t-s+k)\lambda_j}=n\delta_{0,t-s+k}$$

imply

$$\begin{aligned}\frac{2\pi}{n}\sum_{j=1}^{n}e^{ik\lambda_j}I\left(\lambda_j\right)&=\frac{1}{n^2}\sum_{t,s=1}^{n}\left(X_t-\bar{x}\right)\left(X_s-\bar{x}\right)\sum_{j=1}^{n}e^{i(t-s+k)\lambda_j}\\&=\frac{1}{n^2}\sum_{t=1}^{n-k}\left(X_t-\bar{x}\right)\left(X_{t+k}-\bar{x}\right)\cdot n\\&=\hat{\gamma}(k).\end{aligned}$$

Moreover, let for instance $n=2l$. Then

$$I\left(-\lambda_j\right)=I\left(\lambda_j\right),\ I(0)=0$$

implies

$$\begin{aligned}\frac{2\pi}{n}\sum_{j=-m}^{m}e^{ik\lambda_j}I^{0}\left(\lambda_j\right)&=\frac{2\pi}{n}\sum_{j=-m}^{m}\exp\left(-il\frac{2\pi}{2l}j\right)\exp\left(i(k+l)\frac{2\pi}{2l}j\right)I^{0}\left(\lambda_j\right)\\&=\frac{2\pi}{n}\sum_{j=1}^{2l}e^{ik\lambda_j}I^{0}\left(\lambda_j\right)=\frac{2\pi}{n}\sum_{j=1}^{2l}e^{ik\lambda_j}I\left(\lambda_j\right).\end{aligned}$$

Theorem 9.5 *Assumptions:*

$$X_t \in \mathbb{R}\ (t \in \mathbb{Z})\ \text{weakly stationary}$$

$$\sum_{k=-\infty}^{\infty} |\gamma_X(k)| < \infty$$

Then

$$E[\hat{\gamma}(k)] = \gamma_X(k) + O\left(n^{-1}\right).$$

Proof W.l.o.g. $E(X_t) = 0$. Then

$$E[(X_t - \bar{x})(X_{t+k} - \bar{x})] = \gamma_X(k) + E[(-X_t - X_{t+k} + \bar{x})\bar{x}]$$

so that

$$\begin{aligned} E[\hat{\gamma}(k)] &= n^{-1}\sum_{t=1}^{n-k} E[(X_t - \bar{x})(X_{t+k} - \bar{x})] \\ &= \left(1 - \frac{k}{n}\right)\gamma_X(k) + \left(1 - \frac{k}{n}\right) var(\bar{x}) + r_n \\ &= \gamma_X(k) + O\left(n^{-1}\right) + r_n \end{aligned}$$

with

$$\begin{aligned} |r_n| &\le n^{-2}\left[\sum_{s=1}^{n}\sum_{t=1}^{n} |\gamma_X(t-s)| + \sum_{s=1}^{n}\sum_{t=1}^{n} |\gamma_X(t+k-s)|\right] \\ &= n^{-1}\left[\sum_{u=-(n-1)}^{n-1} |\gamma_X(u)| + \sum_{u=-(n-1)}^{n-1} |\gamma_X(u+k)|\right] \\ &= O\left(n^{-1}\right). \end{aligned}$$

Remark 9.5 Theorem 9.5 is not applicable for

$$\frac{k_n}{n} \to \eta > 0$$

Remark 9.6 Could one use

$$\tilde{\gamma}(k) = \frac{1}{n-k}\sum_{t=1}^{n-k} (X_t - \bar{x})(X_{t+k} - \bar{x}),$$

to reduce the bias? No, in general $\tilde{\gamma}$ is not positive semidefinite.

Example 9.3 For

$$n = 2, X_1 = -1, X_2 = 1,$$

$$\tilde{\gamma}(0) = 1, \tilde{\gamma}(\pm 1) = -1, \tilde{\gamma}(k) = 0\ (|k| \geq 2)$$

Thus, $\tilde{\gamma}$ is not positive semidefinite.

Theorem 9.6 *Assumptions:*

$$\varepsilon_t \in \mathbb{R}\ (t \in \mathbb{Z})\ iid$$

$$E(\varepsilon_t) = 0, \sigma_\varepsilon^2 = var(\varepsilon_t) < \infty, E\left(\varepsilon_t^4\right) = \kappa\sigma_\varepsilon^4 < \infty$$

$$X_t = \mu + \sum_{j=-\infty}^{\infty} a_j \varepsilon_{t-j}\ (t \in \mathbb{Z}),\ acf\ \gamma_X$$

$$\sum_{j=-\infty}^{\infty} |a_j| < \infty$$

Then, for fixed $k, l \in \mathbb{N}$,

$$\lim_{n\to\infty} ncov(\hat{\gamma}(k), \hat{\gamma}(l))$$

$$= (\kappa - 3)\gamma_X(k)\gamma_X(l) + \sum_{u=-\infty}^{\infty} [\gamma_X(u)\gamma_X(u + l - k) + \gamma_X(u + k)\gamma_X(u - l)]$$

Proof See e.g. Brockwell and Davis (1991, Proposition 7.3.1).

Corollary 9.4 *Assumptions:*

$$X_t \in \mathbb{R}\ (t \in \mathbb{Z})\ iid$$

$$\sigma^2 = var(X_t) < \infty, E\left(X_t^4\right) = \kappa\sigma_\varepsilon^4 < \infty$$

Then

$$\lim_{n\to\infty} nvar(\hat{\gamma}(0)) = (\kappa - 1)\sigma^4,$$

$$\lim_{n\to\infty} nvar(\hat{\gamma}(k)) = \sigma^4\ (k \neq 0),$$

$$\lim_{n\to\infty} ncov(\hat{\gamma}(k), \hat{\gamma}(l)) = 0\ (k \neq l).$$

Remark 9.7 Corollary 9.4 is not applicable to uncorrelated processes in general, because it assumes independence (see e.g. Basrak et al. 1999; Mikosch and Starica 2000).

Remark 9.8 For the case with $\sum_{k=-\infty}^{\infty}|\gamma_X(k)| = \infty$ see e.g. Beran et al. (2013, Chapter 4.4) and references therein.

Notation 25

$$\hat{\rho}_{1,m} = \left(\hat{\rho}(1)\cdots\hat{\rho}(m)\right)^T, \rho_{1,m} = \left(\rho(1)\cdots\rho(m)\right)^T$$

$$\alpha_{ij}(k) = \rho(k+i)\rho(k+j) + \rho(k-i)\rho(k+j)$$

$$\beta_{ij}(k) = 2\rho(k)\left[\rho(i)\rho(j)\rho(k) - \rho(i)\rho(k+j) - \rho(j)\rho(k+i)\right]$$

$$v_{ij} = v_{ij}(k) = \sum_{k=-\infty}^{\infty}\alpha_{ij}(k) + \sum_{k=-\infty}^{\infty}\beta_{ik}(k) \tag{9.5}$$

$$\underset{m\times m}{V} = V(k) = \left(v_{ij}\right)_{i,j=1,\ldots,m}$$

Theorem 9.7 *Assumptions:*

$$\varepsilon_t \in \mathbb{R}\ (t\in\mathbb{Z})\ iid$$

$$E(\varepsilon_t) = 0, \sigma_\varepsilon^2 = var(\varepsilon_t) < \infty, E\left(\varepsilon_t^4\right) = \kappa\sigma_\varepsilon^4 < \infty$$

$$X_t = \mu + \sum_{j=-\infty}^{\infty} a_j\varepsilon_{t-j}\ (t\in\mathbb{Z}),\ acf\ \gamma_X$$

$$\sum_{j=-\infty}^{\infty}\left|a_j\right| < \infty$$

Then

$$\sqrt{n}\left(\hat{\rho}_{1,m} - \rho_{1,m}\right) \underset{d}{\rightarrow} N(0, V)$$

Proof See e.g. Brockwell and Davis (1991).

Remark 9.9 Equation (9.5) is known as Bartlett's formula.

Corollary 9.5 *Assumptions:*

$$X_t \in \mathbb{R}\ (t\in\mathbb{Z})\ iid$$

$$\sigma^2 = var(X_t) < \infty, E\left(X_t^4\right) = \kappa\sigma_\varepsilon^4 < \infty.$$

Then

$$V = I = diag\,(1, \ldots, 1)\,.$$

Remark 9.10 Corollary 9.5 can be used for testing

$$H_o : X_t \text{ iid vs } H_1 : \rho\,(k) \neq 0.$$

Rejection region:

$$K_\alpha = \left\{ (x_1, \ldots, x_n)^T \in \mathbb{R}^n : |\hat{\rho}\,(k)| > \frac{z_{1-\alpha/2}}{\sqrt{n}} \right\}$$

where

$$z_{1-\alpha/2} = \Phi^{-1}\left(1 - \frac{\alpha}{2}\right).$$

For instance, for $\alpha = 0.05$,

$$K_{0.05} \approx \left\{ (x_1, \ldots, x_n)^T \in \mathbb{R}^n : |\hat{\rho}\,(k)| > \frac{2}{\sqrt{n}} \right\}.$$

Note that K_α is valid for one fixed k only. Moreover,

$$\rho\,(k) = 0 \nRightarrow X_t\ (t \in \mathbb{Z}) \text{ iid}$$

Remark 9.11 $\hat{\rho}(1), \ldots, \hat{\rho}(n-1)$ are linearly dependent:

$$\sum_{k=1}^{n-1} \hat{\rho}\,(k) = \frac{1}{2} \sum_{k=-(n-1)}^{n-1} \hat{\rho}\,(k) - \frac{1}{2} = \pi \frac{I^0\,(0)}{\hat{\gamma}\,(0)} - \frac{1}{2} = -\frac{1}{2}.$$

Remark 9.12 In general, for dependent X_t, $\hat{\rho}(1), \ldots, \hat{\rho}(m)$ are not uncorrelated even for fixed m.

Conclusion 10 *Visual inspection of the correlogram*

$$\hat{\rho}\,(1)\,, \ldots, \hat{\rho}\,(m_n)\ \ (m_n \to \infty)$$

should take into account multiple testing and correlation between $\hat{\rho}(1), \ldots, \hat{\rho}(n-1)$. *Thus,* m_n *should not be too large, and certainly* $m_n = o(n)$.

9.4 Nonparametric Estimation of *f*

Problem 9.4 Orthogonal decomposition of an observed time series $X_1, \ldots, X_n \in \mathbb{R}$ into periodic components.

Solution 9.4 $X_1, \ldots, X_n \in \mathbb{R}$ is a function

$$X : \{1, \ldots, n\} \to \mathbb{R}, t \to X_t$$

in the Hilbert space

$$L_n^2 = \{h \mid h : \{1, \ldots, n\} \to \mathbb{R}\}$$

with

$$\langle f, h \rangle = \langle f, h \rangle_{L_n^2} = \sum_{t=1}^{T_S} f(t)\, h(t) \;\; (f, h \in L_n^2).$$

Hence, find an orthogonal basis in L_n^2 consisting of periodic functions. A simple choice is

$$e_o \equiv 1, e_j(t) = \sin \lambda_j t, e_{j+m}(t) = \cos \lambda_j t \; (j = 1, 2, \ldots, m = \left[\frac{n-1}{2}\right]),$$

and, in addition for n even,

$$e_{n-1}(t) = \cos \pi t = (-1)^t.$$

Lemma 9.10

$$\langle e_j, e_l \rangle = 0 \; (j \neq l)$$

$$||e_o||^2 = n$$

$$||e_j||^2 = \frac{n}{2} \; (1 \leq j \leq n-2)$$

$$\|e_{n-1}\|^2 = \frac{n}{2} \; (n \text{ odd}), \; \|e_{n-1}\|^2 = n \; (n \text{ even})$$

Proof

1)

$$n\lambda_j = n\frac{2\pi j}{n} = 2\pi j$$

so that

$$\sum_{t=1}^{n} \cos \lambda_j t + i \sum_{t=1}^{n} \sin \lambda_j t = \sum_{i=1}^{n} e^{i\lambda_j t} = e^{i\lambda_j} \frac{1 - e^{in\lambda_j}}{1 - e^{i\lambda_j}} = 0$$

Hence,

$$\langle e_o, e_j \rangle = \sum_{t=1}^{n} \sin \lambda_j t \ (1 \le j \le m)$$

and

$$\langle e_o, e_{j+m} \rangle = \sum_{t=1}^{n} \cos \lambda_j t \ (1 \le j \le m).$$

Also, for $1 \le j \le m$,

$$\langle e_j, \epsilon_{j+m} \rangle = \sum_{t=1}^{n} \sin \lambda_j t \cos \lambda_j t = \frac{1}{2} \sum_{t=1}^{n} \sin 2\lambda_j t = 0.$$

A similar proof applies to

$$\langle e_j, e_l \rangle = \sum_{t=1}^{n} \sin \lambda_j t \sin \lambda_l t = 0 \ (j \ne l)$$

and

$$\langle e_{j+m}, e_{l+m} \rangle = \sum_{t=1}^{n} \cos \lambda_j t \cos \lambda_l t \ (j \ne l).$$

2)

$$||e_{j+m}||^2 = \sum_{t=1}^{n} \left(\cos \lambda_j t\right)^2 = \frac{1}{2} \sum_{t=1}^{n} \left(1 + \cos 2\lambda_j t\right) = \frac{n}{2},$$

$$||e_j|^2 = n - \sum_{t=1}^{n} \left(\cos \lambda_j t\right)^2 = \frac{n}{2}$$

$$||e_0||^2 = \sum_{t=1}^{n} 1^2 = n$$

For n even,

$$||e_{n-1}||^2 = \sum_{t=1}^{n} (-1)^2 = n.$$

Corollary 9.6 *Assumptions:*

$$X_1, \ldots, X_n \in \mathbb{R}$$

$$\beta_j = \frac{\langle X, e_j \rangle}{\left\| e_j \right\|^2} = \frac{\sum_{t=1}^{n} X_t e_j(t)}{\sum_{t=1}^{n} e_j^2(t)} \; (j = 0, \ldots, n-1)$$

Then

$$X_t = \sum_{j=0}^{n-1} \beta_j e_j(t) \; (t = 1, \ldots, n)$$

and

$$SST = \sum_{t=1}^{n} (X_t - \bar{x})^2 = \sum_{j=1}^{n-1} \beta_j^2 ||e_j||^2.$$

Proof

$$e_j \; (j = 0, \ldots, n-1) \text{ orthogonal basis}$$

$$\Rightarrow X_t = \sum_{j=0}^{n-1} \beta_j e_j(t), \; \beta_j = \frac{\langle X, e_j \rangle}{\left\| e_j \right\|^2}.$$

$$\Rightarrow \beta_0 = \bar{x},$$

$$\Rightarrow \|X\|^2 = \sum_{t=1}^{n} X_t^2 = \sum_{j=0}^{n-1} \beta_j^2 \left\| e_j \right\|^2 = \bar{x}^2 n + \sum_{j=1}^{n-1} \beta_j^2 \left\| e_j \right\|^2$$

$$\Rightarrow \sum_{t=1}^{n} (X_t - \bar{x})^2 = \sum_{t=1}^{n} X_t^2 - \bar{x}^2 n = \sum_{j=1}^{n-1} \beta_j^2 \left\| e_j \right\|^2$$

Corollary 9.7

$$SST = 4\pi \sum_{j=1}^{m} I\left(\lambda_j\right) \text{ (n odd)},$$

$$SST = 4\pi \sum_{j=1}^{m} I\left(\lambda_j\right) + 2\pi I\left(\pi\right) \text{ (n even)}$$

Proof For

$$n = 2m + 1$$

$$SST = \sum_{j=1}^{m} \left(\beta_j^2 ||e_j||^2 + \beta_{j+m}^2 ||e_{j+m}||^2\right) = \frac{n}{2} \sum_{j=1}^{m} \left(\beta_j^2 + \beta_{j+m}^2\right),$$

$$\beta_j = \frac{2}{n} \sum_{t=1}^{n} (X_t - \bar{x}) \sin \lambda_j t = \frac{2}{n} \sum_{t=1}^{n} X_t \sin \lambda_j t$$

$$\beta_{j+m} = \frac{2}{n} \sum_{t=1}^{n} (X_t - \bar{x}) \cos \lambda_j t = \frac{2}{n} \sum_{t=1}^{n} X_t \cos \lambda_j t$$

$$\begin{aligned} I(\lambda_j) &= \frac{1}{2\pi n} \left\{ \left(\sum_{t=1}^{n} X_t \sin \lambda_j t \right)^2 + \left(\sum_{t=1}^{n} X_t \cos \lambda_j t \right)^2 \right\} \\ &= \frac{1}{2\pi n} \left\{ \left(\frac{n}{2}\right)^2 \beta_j^2 + \left(\frac{n}{2}\right)^2 \beta_{j+m}^2 \right\} = \frac{1}{4\pi} \left\{ \frac{n}{2} \left(\beta_j^2 + \beta_{j+m}^2\right) \right\}. \end{aligned}$$

Hence

$$SST = \frac{n}{2} \sum_{j=1}^{m} \left(\beta_j^2 + \beta_{j+m}^2\right) = 4\pi \sum_{j=1}^{m} I(\lambda_j).$$

For

$$n = 2m,$$

we have to add

$$\beta_{n-1}^2 ||e_{n-1}||^2 = \frac{1}{n} \left(\sum_{t=1}^{n} X_t \cos \pi t \right)^2 = 2\pi I\left(\pi\right).$$

Problem 9.5 Can $I(\lambda)$ be used for estimating $f_X(\lambda)$?

Solution 9.5 $I(\lambda)$ not consistent, but consistency can be achieved by smoothing.

Definition 9.9 Let D_k $(k \in \mathbb{N})$ denote the Dirichlet kernel (see Definition 3.7). Then the sequence of functions

$$H_n : \mathbb{R} \to \mathbb{R}, \lambda \to H_n(\lambda) = \frac{1}{n}\sum_{k=0}^{n-1} D_k(x) \quad (n \in \mathbb{N})$$

is called a Fejér kernel.

Lemma 9.11

$$H_n(x) = \sum_{j=-n}^{n}\left(1 - \frac{|j|}{n}\right)e^{ijx} = \frac{1}{n}\left[\frac{\sin\left(\frac{nx}{2}\right)}{\sin\frac{x}{2}}\right]^2$$

Proof Definition and application of a geometries series and its derivative.

Theorem 9.8 *Assumptions:*

$$X_t \textit{ weakly stationary}$$

$$E(X_t) = \mu, \textit{ acf } \gamma_X, \textit{ spectral density } f_X$$

Then

$$\begin{aligned}
E[I(\lambda)] &= \frac{1}{2\pi}\sum_{k=-(n-1)}^{n-1}\left(1 - \frac{|k|}{n}\right)\gamma(k)\,e^{-ik\lambda} + \mu^2\frac{1}{2\pi}H_n(\lambda) \\
&= \frac{1}{2\pi}\int_{-\pi}^{\pi} H_n(\lambda - \nu)f(\nu)\,d\nu + \mu^2\frac{1}{2\pi}H_n(\lambda)
\end{aligned}$$

Proof

$$\begin{aligned}
E[I(\lambda)] &= \frac{1}{2\pi n}\sum_{t,s=1}^{n} E(X_t X_s)\,e^{-i(t-s)\lambda} = \frac{1}{2\pi n}\sum_{t,s=1}^{n}\left[\gamma_X(t-s) + \mu^2\right]e^{-i(t-s)\lambda} \\
&= \frac{1}{2\pi}\sum_{k=-(n-1)}^{n-1}\left(1 - \frac{|k|}{n}\right)\gamma_X(k)e^{ik\lambda} + \mu^2\frac{1}{2\pi}\sum_{k=-(n-1)}^{n-1}\left(1 - \frac{|k|}{n}\right)e^{ik\lambda}
\end{aligned}$$

Now

$$\frac{1}{2\pi}\sum_{k=-(n-1)}^{n-1}\left(1-\frac{|k|}{n}\right)\gamma(k)e^{ik\lambda}=\frac{1}{2\pi}\int_{-\pi}^{\pi}f(\nu)\sum_{k=-(n-1)}^{n-1}\left(1-\frac{|k|}{n}\right)e^{ik(\lambda-\nu)}d\nu$$
$$=\frac{1}{2\pi}\int_{-\pi}^{\pi}H_n(\lambda-\nu)f(\nu)d\nu$$

and

$$\mu^2\frac{1}{2\pi}\sum_{k=-(n-1)}^{n-1}\left(1-\frac{|k|}{n}\right)e^{ik\lambda}=\mu^2H_n(\lambda).$$

Corollary 9.8 *Assumptions:*

$$X_t \text{ weakly stationary}$$

$$E(X_t)=\mu,\ acf\ \gamma_X,\ spectral\ density\ f_X$$

$$\sum_{k=-\infty}^{\infty}|k|\,|\gamma_X(k)|<\infty$$

Then

$$\sup_{\lambda\in[-\pi,\pi]}\left|E[I(\lambda)]-f_X(\lambda)-\mu^2\frac{1}{2\pi}H_n(\lambda)\right|=O\left(n^{-1}\right)$$

and

$$\sup_{\lambda\in[-\pi,\pi]}\left|E\left[I^0(\lambda)\right]-f_X(\lambda)\right|=O\left(n^{-1}\right)$$

Proof

$$E[I(\lambda)]=\frac{1}{2\pi}\sum_{k=-(n-1)}^{n-1}e^{-ik\lambda}\left(1-\frac{|k|}{n}\right)\left[\gamma_X(k)+\mu^2\right]$$
$$=f(\lambda)+\mu^2\frac{1}{2\pi}H_n(\lambda)+r_{n,1}(\lambda)+r_{n,2}(\lambda)$$

with

$$r_{n,1}(\lambda)=\frac{1}{2\pi}\sum_{|k|\geq n}\gamma_X(k)e^{-ik\lambda},\ r_{n,2}(\lambda)=-\frac{1}{2\pi}\sum_{k=-(n-1)}^{n-1}\frac{|k|}{n}\gamma_X(k)e^{-ik\lambda}.$$

The result follows from

$$|nr_{n,1}(\lambda)| \leq \frac{1}{2\pi} \sum_{k \geq n} |k\gamma_X(k)| \to 0$$

and

$$|r_{n,2}(\lambda)| \leq \frac{1}{2\pi n} \sum_{k=-\infty}^{\infty} |k| \, |\gamma_X(k)| \, .$$

Theorem 9.9 *Assumptions:*

$$\varepsilon_t \; (t \in \mathbb{Z}) \; iid, \; E(\varepsilon_t) = 0, \; \sigma_\varepsilon^2 = var(\varepsilon_t) < \infty$$

$$\sum_{j=-\infty}^{\infty} |a_j| < \infty$$

$$X_t = \sum_{j=-\infty}^{\infty} a_j \varepsilon_{t-j}$$

$$f_X(\lambda) > 0 \; (\lambda \in [-\pi, \pi])$$

$$0 < \nu_1 < \ldots < \nu_k < \pi$$

Then

$$(I(\nu_1), \ldots, I(\nu_k)) \underset{d}{\to} (f_X(\nu_1)\xi_1, \ldots, f_X(\nu_k)\xi_k)$$

with

$$\xi_i \geq 0 \; iid, \; P(\xi_i \leq u) = 1 - \exp(-u) \;\; (u \geq 0)$$

Proof For a detailed proof see e.g. Brockwell and Davis (1991). Some heuristic arguments:

a) Case 1:

$$X_t \text{ iid } N\left(0, \sigma_X^2\right), \; \nu = \frac{2\pi l}{n}$$

Define

$$A(\nu) = \frac{1}{\sqrt{2\pi n}} \sum_{t=1}^{n} X_t \cos \nu t, \; B(\nu) = \frac{1}{\sqrt{2\pi n}} \sum_{t=1}^{n} X_t \sin \nu t$$

Then

$$(A(\nu), B(\nu)) \text{ bivariate Normal,}$$

$$E(A) = E(B) = 0,$$

$$\begin{aligned} var(A) &= var(B) \\ &= \frac{\sigma_X^2}{2\pi n} \sum_{t=1}^{n} \cos^2 \nu t \\ &= \frac{1}{2} \frac{\sigma_X^2}{2\pi} = \frac{1}{2} f_X(\nu) \end{aligned}$$

and

$$cov(A, B) = \frac{\sigma_X^2}{2\pi} \sum_{t=1}^{n} \cos \nu t \sin \nu t = 0.$$

Hence,

$$(A(\nu), B(\nu)) = \sqrt{\frac{1}{2} f_X(\nu)} (Z_1, Z_2)$$

with

$$Z_i \text{ iid } N(0, 1)$$

and

$$I(\nu) = A^2(\nu) + B^2(\nu) = \frac{1}{2} f_X(\nu) \left(Z_1^2 + Z_2^2\right) \sim \frac{1}{2} f_X(\nu) \chi_2^2. \tag{9.6}$$

b) Case 2:

$$X_t \text{ iid } N\left(0, \sigma_X^2\right), \nu \in [-\pi, \pi]$$

Then (9.6) follows from a), because

$$\left\{\frac{2\pi l}{n}, 1 \le l \le \frac{n}{2}, n \in \mathbb{N}\right\} \text{ dense in } [-\pi, \pi].$$

c) Case 3:

$$X_t \text{ iid}, \sigma_X^2 = var(X_t) < \infty.$$

Then (9.6) follows from the CLT,

$$(A(\nu), B(\nu)) \underset{d}{\rightarrow} \sqrt{\frac{1}{2} f_X(\nu)} (Z_1, Z_2).$$

d) Case 4:

$$X_t = X_{t,M} = \sum_{j=-M}^{M} a_j \varepsilon_{t-j}.$$

Then (9.6) follows from the CLT for M-dependent processes,

$$(A(\nu), B(\nu)) \underset{d}{\rightarrow} \sqrt{\frac{1}{2} f_X(\nu)} (Z_1, Z_2).$$

e) Case 5: X_t as in the assumptions of Theorem 9.9. Then (9.6) follows from d) using

$$\lim_{M\to\infty} \|X_t - X_{t,M}\|^2 = \lim_{M\to\infty} E\left[(X_t - X_{t,M})^2\right] = 0.$$

Remark 9.13 In Theorem 9.9, $\nu_1, \ldots, \nu_k$ and k are fixed.

Problem 9.6 The number of Fourier frequencies is not fixed, but proportional to n.

Solution 9.6 For Fourier frequencies, $I(\lambda_j)$ and $I(\lambda_l)$ ($j \neq l$) turn out to be asymptotically uncorrelated with a uniform finite sample upper bound.

Theorem 9.10 *Assumptions:*

$$\varepsilon_t \ (t \in \mathbb{Z}) \ iid, \ E(\varepsilon_t) = 0, \ \sigma_\varepsilon^2 = var(\varepsilon_t) < \infty, \ E(\varepsilon_t^4) = \kappa \sigma_\varepsilon^4 < \infty$$

$$\sum |a_j| \sqrt{|j|} < \infty$$

$$X_t = \sum_{j=-\infty}^{\infty} a_j \varepsilon_{t-j}$$

$$\lambda_j = \frac{2\pi j}{n} \in (0, \pi)$$

Then

$$\exists C_1, C_2 < \infty \ s.t.:$$

$$\left|cov\left(I(\lambda_j), I(\lambda_l)\right)\right| \leq C_1 n^{-1} \ (\lambda_j \neq \lambda_l)$$

and

$$var\left(I\left(\lambda_l\right)\right) = f_X^2\left(\lambda_l\right) + r_n\left(\lambda_l\right)$$

with

$$\left|r_n\left(\lambda_l\right)\right| \leq C_2 n^{-\frac{1}{2}}.$$

Proof See e.g. Brockwell and Davis (1991).

Problem 9.7

$$I\left(\lambda\right) \underset{p}{\nrightarrow} f_X\left(\lambda\right)$$

Solution 9.7 (a) use only $\hat{\gamma}(k)$ ($|k| \leq M_n$) with $M_n/n \to 0$ (lag-window estimator); (b) average $I(\lambda)$ locally (spectral window estimator, smoothed periodogram)

Definition 9.10 Let

$$w_n : \mathbb{Z} \to \mathbb{R}\ (n \in \mathbb{N}) \text{ s.t.:}$$

$$\lim_{n\to\infty} w_n\left(k\right) = 1\ (\forall k \in \mathbb{Z}),\ \lim_{n\to\infty} w_n\left(qn\right) = 0\ (\forall q > 0) \tag{9.7}$$

Then

$$\hat{f}_X\left(\lambda\right) = \frac{1}{2\pi} \sum_{k=-(n-1)}^{n-1} w_n\left(k\right)\hat{\gamma}\left(k\right)e^{ik\lambda} \tag{9.8}$$

is called a lag-window estimator of f_X.

Example 9.4

$$w_n(k) = 1\left\{-M_n \leq k \leq M_n\right\},\ M_n \to \infty,\ \frac{M_n}{n} \to 0$$

$$\Rightarrow \hat{f}\left(\lambda\right) = \frac{1}{2\pi} \sum_{k=-M_n}^{M_n} \hat{\gamma}\left(k\right)e^{ik\lambda}$$

Theorem 9.11 *Assumptions:*

$$\hat{f}\left(\lambda\right) = \frac{1}{2\pi} \sum_{k=-(n-1)}^{n-1} w_n\left(k\right)\hat{\gamma}\left(k\right)e^{-ik\lambda}$$

$$W_n\left(u\right) = \frac{1}{2\pi} \sum_{k=-(n-1)}^{n-1} e^{iku} w_n\left(k\right)$$

Then

$$\hat{f}(\lambda) = \int_{-\pi}^{\pi} W_n(\nu - \lambda) I^0(\nu) d\nu,$$

$$w_n(k) = \int_{-\pi}^{\pi} e^{-iku} W_n(u)\, du$$

and

$$\int_{-\pi}^{\pi} W_n^2(u)\, du = \frac{1}{2\pi} \sum_{k=-(n-1)}^{n-1} w_n^2(k).$$

Proof

$$\begin{aligned}
\hat{f}(\lambda) &= \frac{1}{2\pi} \sum_{k=-(n-1)}^{n-1} w_n(k)\, \hat{\gamma}(k)\, e^{-ik\lambda} \\
&= \frac{1}{2\pi} \sum_{k=-(n-1)}^{n-1} e^{-ik\lambda} w_n(k) \int_{-\pi}^{\pi} e^{ik\nu} I^0(\nu)\, d\nu \\
&= \int_{-\pi}^{\pi} \left[\frac{1}{2\pi} \sum_{k=-(n-1)}^{n-1} e^{ik(\nu-\lambda)} w_n(k) \right] I^0(\nu)\, d\nu \\
&= \int_{-\pi}^{\pi} W_n(\nu - \lambda) I^0(\nu)\, d\nu.
\end{aligned}$$

The other formulas follow from

$$\int_{-\pi}^{\pi} e^{-iku} W_n(u)\, du = \frac{1}{2\pi} \sum_{l=-(n-1)}^{n-1} w_n(l) \int_{-\pi}^{\pi} e^{i(l-k)u} du = w_n(k)$$

and

$$\begin{aligned}
\int_{-\pi}^{\pi} W_n^2(u)\, du &= \frac{1}{(2\pi)^2} \sum_{k,l=-(n-1)}^{n-1} w_n(k)\, w_n(l) \int_{-\pi}^{\pi} e^{i(k-l)u} du \\
&= \frac{1}{2\pi} \sum_{k=-(n-1)}^{n-1} w_n^2(k).
\end{aligned}$$

Example 9.5

$$w(k) = 1\{-M \le k \le M\}$$

$$\Rightarrow W_n(u) = \frac{1}{2\pi} \sum_{k=-M}^{M} e^{iku} = \frac{1}{2\pi} \frac{\sin(M + \frac{1}{2})u}{\sin \frac{u}{2}} = \frac{1}{2\pi} D_M(u)$$

Theorem 9.11 motivates the following general definition:

Definition 9.11 Let

$$W_n : \mathbb{R} \to \mathbb{R}\ (n \in \mathbb{N}) \text{ s.t.:}$$

$$W_n (\lambda + 2\pi) = W_n (\lambda), W_n (-\lambda) = W_n (\lambda), \lim_{n\to\infty} W_n (u) = 0\ (u \neq 0) \tag{9.9}$$

$$\lim_{n\to\infty} \int_{-\pi}^{\pi} W_n (u)\, du = 1, \lim_{n\to\infty} n^{-1} \int_{-\pi}^{\pi} W_n^2 (u)\, du = 0 \tag{9.10}$$

Then

$$\hat{f} (\lambda) = \int_{-\pi}^{\pi} W_n (\nu - \lambda)\, I^0(\nu) d\nu \tag{9.11}$$

is called a spectral window estimator of f_X.

Related definitions are:

Definition 9.12 Define a kernel function

$$K : \mathbb{R} \to \mathbb{R} \text{ s.t.:}$$

$$K (u) \ge 0, K (-u) = K (u), K (u) = 0\ (|u| > 1)$$

$$\int K (u)\, du = 1, \int K^2 (u)\, du < \infty,$$

a sequence of bandwidths

$$b_n > 0 \text{ s.t. } b_n \to 0, nb_n \to \infty$$

and

$$W_n (u) = b_n^{-1} K \left(\frac{u}{b_n} \right).$$

Then

$$\hat{f}(\lambda) = \int_{-\pi}^{\pi} W_n(\lambda - \nu) I^0(\nu)\, d\nu = b_n^{-1} \int_{-\pi}^{\pi} K\left(\frac{\lambda - \nu}{b_n}\right) I^0(\nu)\, d\nu \tag{9.12}$$

is called a kernel spectral estimator of f. Alternatively, using a Riemann sum approximation and only Fourier frequencies $|\lambda_j| \leq \pi$,

$$\hat{f}(\lambda) = b_n^{-1} \sum_{j=-m}^{m} K\left(\frac{\lambda - \lambda_j}{b_n}\right) I^0(\lambda_j) \frac{2\pi}{n} \tag{9.13}$$

is called a kernel spectral estimator of f.

Definition 9.13 Let W_n be as in Definition 9.11 or 9.12,

$$\lambda_j = \frac{2\pi j}{n}, m = \left[\frac{n-1}{2}\right], M_n \to \infty, \frac{M_n}{n} \to 0$$

$$w_j\ (j \in \mathbb{Z}) \text{ s.t.:}$$

$$w_n(j) \geq 0, w_n(-j) = w_n(j)$$

$$w_n(j) = 0\ (|j| > M_n)$$

$$\sum_{j=-\infty}^{\infty} w_n(j) = \sum_{j=-M_n}^{M_n} w_n(j) = 1$$

$$\sum_{j=-\infty}^{\infty} w_n^2(j) = \sum_{j=-M_n}^{M_n} w_n^2(j) \to 0$$

Then

$$\hat{f}(\lambda_j) = \sum_{l=-m}^{m} w_n(l) I^0(\lambda_{j+l}) \tag{9.14}$$

is called a discrete spectral average estimator of $f(\lambda_j)$. Moreover, for nonzero frequencies $\lambda \notin \{\pm\lambda_j, 1 \leq j \leq m\}$, $\hat{f}(\lambda)$ is defined by $\hat{f}(\lambda_j)$ where λ_j is the Fourier frequency closest to λ.

Theorem 9.12 *Assumptions:*

$$\varepsilon_t \ (t \in \mathbb{Z}) \ iid, \ E(\varepsilon_t) = 0, \ \sigma_\varepsilon^2 = var(\varepsilon_t), \ E(\varepsilon_t^4) < \infty$$

$$\sum_{j=0}^{\infty} |a_j| \sqrt{j} < \infty$$

$$X_t = \sum_{j=0}^{\infty} a_j \varepsilon_{t-j}$$

$$\hat{f}(\lambda) \text{ as in Definition 9.13}$$

Then

$$\lim_{n\to\infty} \sup_{\lambda\in[-\pi,\pi]\setminus\{0\}} \left| E\left[\hat{f}(\lambda)\right] - f(\lambda) \right| = 0$$

$$\lim_{n\to\infty} \frac{var\left(\hat{f}(\lambda)\right)}{\sum_{j=-M_n}^{M_n} w_n^2(j)} = f^2(\lambda)$$

$$\lim_{n\to\infty} \frac{cov\left(\hat{f}(\nu), \hat{f}(\lambda)\right)}{\sum_{j=-M_n}^{M_n} w_n^2(j)} = 0 \ (\nu \neq \lambda)$$

$$\lim_{n\to\infty} \sup_{\lambda\in[-\pi,\pi]\setminus\{0\}} E\left[\left(\hat{f}(\lambda) - f(\lambda)\right)^2\right] = 0$$

Proof See e.g. Brockwell and Davis (1991, Theorem 10.4.1).

Definition 9.14 Let

$$\varepsilon_t \ (t \in \mathbb{Z}) \text{ iid}, \ E(\varepsilon_t) = 0, \ \sigma_\varepsilon^2 = var(\varepsilon_t) < \infty$$

$$\eta_t \ (t \in \mathbb{Z}) \text{ iid}, \ P(\eta_t \le x) = P(\varepsilon_t \le x) \ (x \in \mathbb{R})$$

$$(\varepsilon_t)_{t\in\mathbb{Z}} \text{ independent of } (\eta_t)_{t\in\mathbb{Z}}$$

$$X_t \ (t \in \mathbb{Z}) \text{ weakly stationary}, \ E(X_t^4) < \infty$$

$$X_t = g(\varepsilon_s, s \le t)$$

For $t \geq 0$, $p > 0$, define

$$\delta_{t,p} = \|X_t - g(\varepsilon_t, \varepsilon_{t-1}, \varepsilon_{t-2}, \ldots, \varepsilon_1, \eta_0, \varepsilon_{-1}, \varepsilon_{-2}, \ldots)\|_{L^p(\Omega)}$$
$$= \{E[|X_t - g(\varepsilon_t, \varepsilon_{t-1}, \varepsilon_{t-2}, \ldots, \varepsilon_1, \eta_0, \varepsilon_{-1}, \varepsilon_{-2}, \ldots)|^p]\}^{1/p},$$

and

$$\Theta_{0,p} = \sum_{t=0}^{\infty} \delta_{t,p}.$$

Theorem 9.13 *Assumptions:*

$$\varepsilon_t \ (t \in \mathbb{Z}) \ iid, \ E(\varepsilon_t) = 0, \ \sigma_\varepsilon^2 = var(\varepsilon_t) < \infty$$

$$X_t \ (t \in \mathbb{Z}) \ \text{weakly stationary}, \ E(X_t) = 0, \ acf \ \gamma_X$$

$$X_t = g(\varepsilon_s, s \leq t)$$

$$\Theta_{0,2} < \infty$$

$$\hat{f}_X(\lambda) = \frac{1}{2\pi} \sum_{k=-(n-1)}^{n-1} w_n(k) \hat{\gamma}(k) e^{ik\lambda} = \frac{1}{2\pi} \sum_{k=-(n-1)}^{n-1} K\left(\frac{k}{M_n}\right) \hat{\gamma}(k) e^{ik\lambda}$$

$$K : \mathbb{R} \to \mathbb{R}, \ K \in C(\mathbb{R})$$

$$K(-u) = K(u), \ K(0) = 1$$

$$\sup_{u \in \mathbb{R}} |K(u)| < \infty, \ \int_{-\infty}^{\infty} |K(u)| \, du < \infty$$

$$\int_{-\infty}^{\infty} K^2(u) \, du = \lim_{M \to \infty} M^{-1} \sum_{k=-\infty}^{\infty} K^2\left(\frac{k}{M}\right) < \infty$$

$$\varphi_K(x) = \int_{-\infty}^{\infty} K(u) e^{ixu} du, \ \int_{-\infty}^{\infty} |\varphi_K(x)| \, dx < \infty$$

$$M_n \to \infty, \ \frac{M_n}{n} \to 0$$

Then

$$\lim_{n \to \infty} \sup_{\lambda \in [-\pi, \pi]} E\left[\left|\hat{f}_X(\lambda) - f_X(\lambda)\right|\right] = 0$$

Proof See Liu and Wu (2010) .

Remark 9.14

$$\Theta_{0,2} < \infty \Rightarrow \sum_{k=-\infty}^{\infty} |\gamma_X(k)| < \infty$$

Theorem 9.14 *Assumptions as in Theorem 9.13, and*

$$E\left(X_t^4\right) < \infty$$

$$\Theta_{0,4} < \infty$$

$$I_2(K) = \int_{-\infty}^{\infty} K^2(u)\, du < \infty$$

$$\lim_{c\to\infty} \sup_{1\le M<\infty} M \sum_{k\ge cM}^{\infty} K^2\left(\frac{k}{M}\right) = 0$$

Then

$$\sqrt{nb}\left\{\hat{f}(\lambda) - E\left[\hat{f}(\lambda)\right]\right\} \underset{d}{\to} Z \sim N\left(0, \sigma^2(\lambda)\right)$$

where

$$\sigma^2(\lambda) = f_X^2(\lambda)\, I_2(K) \ \ (\lambda \notin \{-\pi, 0, \pi\}),$$

and

$$\sigma^2(\lambda) = 2f_X^2(\lambda)\, I_2(K) \ \ (\lambda \in \{-\pi, 0, \pi\})$$

Proof See Liu and Wu (2010)

Remark 9.15 For results under more restrictive assumptions see e.g. Brillinger (1969, 2001) and references therein, also Anderson (1971), Priestley (1981), Rosenblatt (1984), Zurbenko (1986), Shao and Wu (2007). In particular, under more restrictive conditions it can be shown that for $\nu \neq \lambda$, $\hat{f}(\nu)$ and $\hat{f}(\lambda)$ are asymptotically independent.

Problem 9.8 Choice of bandwidth b?

Solution 9.8 Minimize asymptotic formula for MSE

Theorem 9.15 *Assumptions:*

$$X_t\ (t\in\mathbb{Z})\text{ weakly stationary, } E\left(X_t\right)=0,\ E\left(X_t^4\right)<\infty$$

$$\text{acf } \gamma_X,\text{ spectral density } f_X\in C^3[-\pi,\pi]$$

$$\sum_{k=-\infty}^{\infty}|k|\left|\gamma_X(k)\right|<\infty$$

$$b\to 0,\ nb^3\to\infty$$

$$K\geq 0,\ K(u)=0\ (|u|>\pi),\ K(-u)=K(u),\ \int K(u)\,du=1$$

$$\exists K'\text{ Lebesgue-a.e.},\ \sup\left|K'(u)\right|<\infty$$

$$\hat{f}(\lambda)=\frac{2\pi}{nb}\sum_{j=-m}^{m}K\left(\frac{\lambda-\lambda_j}{b}\right)I^0\left(\lambda_j\right)$$

Then

$$E[\hat{f}(\lambda)]=f(\lambda)+\frac{1}{2}b^2 f_X''(\lambda)\int_{-\pi}^{\pi}K(u)u^2du+o(b^2)+O\left((nb)^{-1}\right)$$

and

$$var\left(\hat{f}(\lambda)\right)=\frac{2\pi f_X^2(\lambda)}{nb}\int_{-\pi}^{\pi}K^2(x)\,dx+o\left((nb)^{-1}\right)$$

Proof

1) Bias: W.l.o.g. $\mu=0$. Then

$$E[I(\lambda)]=\frac{1}{2\pi}\int_{-\pi}^{\pi}H_n(\lambda-\nu)f(\nu)\,d\nu=f_X(\lambda)+O\left(n^{-1}\right)$$

and

$$\begin{aligned}E\left[\hat{f}(\lambda)\right]&=\frac{2\pi}{nb}\sum K\left(\frac{\lambda-\lambda_j}{b}\right)E\left[I\left(\lambda_j\right)\right]\\&=\frac{2\pi}{nb}\sum K\left(\frac{\lambda-\lambda_j}{b}\right)f_X\left(\lambda_j\right)+O\left(n^{-1}\right).\end{aligned}$$

Now

$$\frac{2\pi}{nb}\sum K\left(\frac{\lambda-\lambda_j}{b}\right)f_X(\lambda_j)$$
$$=\frac{2\pi}{nb}\sum K\left(\frac{\lambda-\lambda_j}{b}\right)\left[f_X(\lambda)+(\lambda_j-\lambda)f_X'(\lambda)+\frac{(\lambda_j-\lambda)^2}{2}f_X''(\lambda)\right.$$
$$\left.+O\left((\lambda_j-\lambda)^3\right)\right].$$

The result follows from

$$\frac{2\pi}{nb}\sum K\left(\frac{\lambda-\lambda_j}{b}\right)=1+O\left((nb)^{-1}\right),$$

$$\frac{2\pi}{nb}\sum K\left(\frac{\lambda-\lambda_j}{b}\right)(\lambda_j-\lambda)=O\left(b\,(nb)^{-1}\right),$$

$$\frac{2\pi}{nb}\sum K\left(\frac{\lambda-\lambda_j}{b}\right)\frac{(\lambda_j-\lambda)^2}{2}=\frac{b^2}{2}\int_{-\pi}^{\pi}u^2K(u)\,du+O\left((nb)^{-1}\right).$$

2) Variance:

$$var\left(\hat{f}(\lambda)\right)=\left(\frac{2\pi}{nb}\right)^2\sum_{j,l}K\left(\frac{\lambda-\lambda_j}{b}\right)K\left(\frac{\lambda-\lambda_l}{b}\right)cov\left(I(\lambda_j),I(\lambda_l)\right)$$
$$=\left(\frac{2\pi}{nb}\right)^2\sum_j\left\{K^2\left(\frac{\lambda-\lambda_j}{b}\right)\left[var\left(I(\lambda_j)\right)+O\left(n^{-1}\right)\right]\right\}$$
$$=\left(\frac{2\pi}{nb}\right)^2\sum_j\left\{K^2\left(\frac{\lambda-\lambda_j}{b}\right)\left[f_X^2(\lambda_j)+O\left(n^{-1}\right)\right]\right\}.$$

The result then follows from

$$\left(\frac{2\pi}{nb}\right)^2\sum_j K^2\left(\frac{\lambda-\lambda_j}{b}\right)f_X^2(\lambda_j)=\frac{2\pi f_X^2(\lambda)}{nb}\sum_j K^2\left(\frac{\lambda-\lambda_j}{b}\right)\frac{2\pi}{nb}+o\left((nb)^{-1}\right)$$
$$=\frac{2\pi f_X^2(\lambda)}{nb}\left(\int_{-\pi}^{\pi}K^2(x)\,dx\right.$$
$$\left.+O\left((nb)^{-1}\right)\right)+o\left((nb)^{-1}\right)$$
$$=\frac{2\pi f_X^2(\lambda)}{nb}\int_{-\pi}^{\pi}K^2(x)\,dx+o\left((nb)^{-1}\right)$$

Corollary 9.9 *Assumptions as in Theorem 9.15. Then*

$$\begin{aligned} MSE &= E\left[\left(\hat{f}(\lambda) - f_X(\lambda)\right)^2\right] \\ &= C_1 b^4 + C_2 (nb)^{-1} + o\left(\max\left(b^4, (nb)^{-1}\right)\right) \end{aligned}$$

where

$$C_1 = \frac{1}{4}\left[f_X''(\lambda)\right]^2, \; C_2 = 2\pi f_X^2(\lambda)\int_{-\pi}^{\pi} K^2(x)\,dx.$$

Moreover, MSE is minimized by

$$b_{opt} = \left[\frac{C_2}{4C_1}\right]^{\frac{1}{5}} n^{-\frac{1}{5}}. \tag{9.15}$$

Remark 9.16 Equation (9.15) depends on unknown quantities. For data driven estimation of b_{opt} see e.g. Beltrao and Bloomfield (1987) and Franke and Härdle (1992).

Remark 9.17 Similar results for the periodogram and its smoothed versions hold for multivariate time series (see e.g. Hannan 1970; Brockwell and Davis 1991, Ch. 11.7).

Chapter 10
Parametric Estimation

10.1 Gaussian and Quasi Maximum Likelihood Estimation

Problem 10.1 Linear process

$$X_t = \sum_{j=0}^{\infty} a_j \varepsilon_{t-j} = A(B)\,\varepsilon_t \tag{10.1}$$

$$\sum_{j=0}^{\infty} a_j^2 < \infty \tag{10.2}$$

$$\varepsilon_t\ (t \in \mathbb{Z}) \text{ iid, } E(\varepsilon_t) = 0,\ \sigma_\varepsilon^2 = var(\varepsilon_t) < \infty \tag{10.3}$$

$$F_{\varepsilon,1}(x) = P\left(\frac{\varepsilon_t}{\sigma_\varepsilon} \le x\right),\ F_{\varepsilon,1} \text{ known} \tag{10.4}$$

$$\text{spectral density } f_X(\lambda;\theta) = \frac{\sigma_\varepsilon^2}{2\pi}\left|\sum_{j=0}^{\infty} a_j e^{-ij\lambda}\right|^2$$

$$a_j = a_j(\eta),\ \eta \in \mathbb{R}^p$$

$$\theta = \left(\sigma_\varepsilon^2, \eta\right) \in \Theta \subseteq \mathbb{R}^{p+1} \text{ unknown}$$

$$\Rightarrow \text{estimation of } \theta?$$

J. Beran, *Mathematical Foundations of Time Series Analysis*,
https://doi.org/10.1007/978-3-319-74380-6_10

Solution 10.1 Gaussian maximum likelihood estimation: (a) simple; (b) optimal for Gaussian processes; (c) under mild moment conditions, consistency and CLT hold even if F_ε is not normal.

Remark 10.1 The distribution of X_t $(t \in \mathbb{Z})$ is fully determined by $F_{\varepsilon,1}$ and θ.

Lemma 10.1 *Assumptions:*

$$X_t \ (t \in \mathbb{Z}) \text{ purely stochastic stationary Gaussian process}$$

$$E(X_t) = 0$$

$$\text{spectral density } f_X(\lambda; \theta) \ \ (\lambda \in [-\pi, \pi]\,), \text{ acf } \gamma_X(k; \theta) \ \ (k \in \mathbb{Z})$$

$$\Sigma_n(\theta) = [\gamma_X(t-s)]_{s,t=1,\ldots,n}$$

Then

$$P(X_1 \le x_1, \ldots, X_n \le x_n) = \int_{-\infty}^{x_1} \cdots \int_{-\infty}^{x_n} p_X(\mathbf{y}; \theta)\, dy_1 \cdots dy_n$$

where

$$\mathbf{y} = (y_1, .., y_n)^T,$$

$$p_X(\mathbf{y}; \theta) = (2\pi)^{-\frac{n}{2}} |\Sigma_n|^{-\frac{1}{2}} \exp\left(-\frac{1}{2}\mathbf{y}^T \Sigma_n^{-1} \mathbf{y}\right) \ \ (\mathbf{y} \in \mathbb{R}^n)$$

and

$$|\Sigma_n| = \det \Sigma_n.$$

Proof Follows from the definition of purely stochastic Gaussian processes and Gaussian multivariate distributions.

Lemma 10.1 leads to the following definition:

Definition 10.1 Let

$$\theta = \left(\sigma_\varepsilon^2, \eta\right) \in \Theta \subseteq \mathbb{R}^{p+1},$$

$$f_X(\lambda; \theta) \ \ (\lambda \in [-\pi, \pi]\,, \theta \in \Theta) \text{ family of spectral densities},$$

$$\Sigma_n(\theta) = [\gamma_X(t-s; \theta)]_{s,t=1,\ldots,n},$$

$$\mathbf{x} = (x_1, \ldots, x_n)^T \in \mathbb{R}^n.$$

Suppose that (10.1)–(10.4) hold with

$$X_t = \sum_{j=0}^{\infty} a_j\left(\eta^0\right)\varepsilon_{t-j} = A(B)\,\varepsilon_t,$$

$$\theta^0 \in \Theta^0,$$

$$F_{\varepsilon,1}(x) = \Phi(x),$$

$$\mathbf{x} = (x_1, \ldots, x_n)^T = (X_1(\omega), \ldots, X_n(\omega))^T.$$

Then

$$\mathscr{L}_n(\theta) = \mathscr{L}_n(\theta; X_1, \ldots, X_n) = \frac{1}{n}\log|\Sigma_n(\theta)| + \frac{1}{n}\mathbf{x}^T\Sigma_n^{-1}(\theta)\,\mathbf{x}\ (\theta \in \Theta) \tag{10.5}$$

is the loglikelihood function, and

$$\hat{\theta} = \operatorname{argmin} \mathscr{L}_n(\theta) \tag{10.6}$$

the maximum likelihood estimator, or MLE, of θ^0. If

$$F_{\varepsilon,1}(x) \neq \Phi(x),$$

then $\mathscr{L}_n(\theta)$ is called a quasi-loglikelihood function and $\hat{\theta}$ a quasi maximum likelihood estimator, or QMLE, of θ^0.

Theorem 10.1 *Assumptions (10.1)–(10.4) with parameter θ^0, and*

$$\theta^0 \in \Theta^0$$

$$E\left(\varepsilon_t^4\right) = \kappa\sigma_\varepsilon^4$$

$$\exists d \in \left(-\frac{1}{2}, \frac{1}{2}\right)\ s.t.\ f_X\left(\lambda; \theta^0\right) \sim c_f|\lambda|^{-2d}\ (\lambda \to 0)$$

$$\theta^0 = \left(\sigma_\varepsilon^2, \eta\right) \in \Theta \subseteq \mathbb{R}^{p+1}$$

$$\hat{\theta} = argmin\ \mathscr{L}_n(\theta)$$

Then, under additional regularity conditions,

$$\sqrt{n}\left(\hat{\theta} - \theta^0\right) \underset{d}{\rightarrow} N(0, \Sigma_{\mathrm{MLE}}) \tag{10.7}$$

where

$$\Sigma_{MLE} = 4\pi C_{\mathrm{MLE}}^{-1} + \kappa\sigma_\varepsilon^4 J,$$

$$J = [J_{rs}], J_{rs} = 1\{r = s = 1\}$$

and

$$C_{\mathrm{MLE}} = [c_{jl}]_{j,l=1,\ldots,p+1}$$

with

$$c_{jl} = \int_{-\pi}^{\pi} \left(\frac{\partial}{\partial\theta_j} \log f_X(\lambda\,\theta^0)\right)\left(\frac{\partial}{\partial\theta_l} \log f_X(\lambda\,\theta^0)\right) d\lambda \mid_{\theta=\theta^0}$$

Proof See e.g. Yajima (1985), Dahlhaus (1989), Hosoya (1997) (also see Beran et al. 2013, Chapter 5.5).

Remark 10.2 Additional regularity conditions include in particular: (a) $f(\lambda)$ ($\lambda \neq 0$) continuous in λ; (b) $\int \log f_X(\lambda;\theta)d\lambda$ twice differentiable w.r.t. θ, and differentiation and integration may be interchanged.

10.2 Whittle Approximation

Problem 10.2 Exact MLE and QMLE complicated theoretically and numerically.

Solution 10.2 Asymptotically equivalent approximations.

Notation 26

$$f_X(\lambda;\theta) = \frac{\sigma_\varepsilon^2}{2\pi}\left|A\left(e^{-i\lambda};\eta\right)\right|^2 = \frac{\sigma_\varepsilon^2}{2\pi} h_X(\lambda;\eta)$$

$$\hat{h}_k = \frac{1}{2\pi}\int_{-\pi}^{\pi} e^{ik\lambda} h_X(\lambda;\eta)\,d\lambda$$

$$\gamma_X(k;\theta) = cov(X_t, X_{t+k}) = \frac{\sigma_\varepsilon^2}{2\pi}\int_{-\pi}^{\pi} e^{ik\lambda} h_X(\lambda;\eta)\,d\lambda$$

$$\Sigma_n(\theta) = [\gamma_X(j-l;\theta)]_{j,l=1,\ldots,n},\ \Sigma_{h,n}(\eta) = 2\pi\left[\hat{h}_{j-l}\right]_{j,l=1,\ldots,n}$$

Remark 10.3

$$\Sigma_n(\theta) = \frac{\sigma_\varepsilon^2}{2\pi}\Sigma_{h,n}(\eta),\ |\Sigma_n(\theta)| = \left(\frac{\sigma_\varepsilon^2}{2\pi}\right)^n |\Sigma_{h,n}(\eta)|$$

Lemma 10.2

$$\mathscr{L}_n(\theta) = \log\frac{\sigma_\varepsilon^2}{2\pi} + \frac{1}{n}\log|\Sigma_{h,n}(\eta)| + \left(\frac{2\pi}{\sigma_\varepsilon^2}\right)\frac{1}{n}\mathbf{x}^T\Sigma_{h,n}^{-1}(\eta)\,\mathbf{x}$$

Proof Follows from

$$\Sigma_n(\theta) = \frac{\sigma_\varepsilon^2}{2\pi}\Sigma_{h,n}(\eta),\ |\Sigma_n(\theta)| = \left(\frac{\sigma_\varepsilon^2}{2\pi}\right)^n |\Sigma_{h,n}(\eta)|$$

Lemma 10.3 *Under suitable regularity conditions,*

$$\lim_{n\to\infty}\frac{1}{n}\log|\Sigma_{h,n}(\eta)| = \frac{1}{2\pi}\int_{-\pi}^{\pi}\log h_X(\lambda;\eta)\,d\lambda$$

Proof See Grenander and Szegö (1958).

Corollary 10.1

$$\lim_{n\to\infty}\frac{1}{n}\log|\Sigma_{h,n}(\theta)| = 0$$

Proof Follows from Lemma 10.3 and (8.4).

Thus, we replace $\mathscr{L}_n$ in (10.5) by the approximation

$$\mathscr{L}_{n,\text{approx}}(\theta) = \int_{-\pi}^{\pi}\log f_X(\lambda;\theta)\,d\lambda + \left(\frac{2\pi}{\sigma_\varepsilon^2}\right)\frac{1}{n}\mathbf{x}^T\Sigma_{h,n}^{-1}(\eta)\,\mathbf{x} \tag{10.8}$$

$$= \log\sigma_\varepsilon^2 + \left(\frac{2\pi}{\sigma_\varepsilon^2}\right)\frac{1}{n}\mathbf{x}^T\Sigma_{h,n}^{-1}(\eta)\,\mathbf{x}. \tag{10.9}$$

Then minimization w.r.t. η can be done separately,

$$\hat{\eta} = \hat{\eta}_{n,\text{approx MLE}} = \underset{\eta}{\operatorname{argmin}}\frac{1}{n}\mathbf{x}^T\Sigma_{h,n}^{-1}(\eta)\,\mathbf{x}, \tag{10.10}$$

and

$$\hat{\sigma}_\varepsilon^2 = \frac{2\pi}{n}\mathbf{x}^T\Sigma_{h,n}^{-1}(\hat{\eta})\,\mathbf{x}. \tag{10.11}$$

Problem 10.3 Calculation of $\Sigma_{h,n}^{-1}$ in (10.10) and (10.11) numerically cumbersome.

Solution 10.3 Whittle approximation, $\Sigma_{h,n}^{-1}$ is replaced by

$$\Sigma_{h,n}^{-1}(\eta) \approx W_n(\eta) = \left[w_{j-l}(\eta)\right]_{j,l=1,\ldots,n} \tag{10.12}$$

where

$$w_{j-l}(\eta) = \frac{1}{(2\pi)^2}\int_{-\pi}^{\pi} e^{i(j-l)\lambda}\frac{1}{h_X(\lambda;\eta)}d\lambda. \tag{10.13}$$

Definition 10.2 Let W_n be defined as in (10.12) and (10.13), and define

$$Q_{n,\text{Whittle}}(\eta) = \mathbf{x}^T\frac{\partial}{\partial\eta}W_n(\eta)\,\mathbf{x}.$$

Then

$$\mathcal{L}_{n,\text{Whittle}}(\theta) = \log\sigma_\varepsilon^2 + \left(\frac{2\pi}{\sigma_\varepsilon^2}\right)\frac{1}{n}\mathbf{x}^T W_n(\eta)\,\mathbf{x} \tag{10.14}$$

is called Whittle loglikelihood, and $\hat{\theta} = \hat{\theta}_{Whittle}$ obtained by

$$Q_{n,\text{Whittle}}(\hat{\eta}) = 0 \tag{10.15}$$

and

$$\hat{\sigma}_\varepsilon^2 = \frac{2\pi}{n}\mathbf{x}^T W_n(\hat{\eta})\,\mathbf{x} = \frac{2\pi}{n}\sum_{j,l=1}^{n} w_{j-l}X_tX_s$$

is called Whittle estimator of θ^0.

Theorem 10.2 *Under the assumptions of Theorem 10.1 and additional regularity assumptions,*

$$\sqrt{n}\left(\hat{\theta}_{Whittle} - \theta^0\right) \underset{d}{\rightarrow} N(0, \Sigma_{\text{MLE}})$$

where Σ_{MLE} is given in (10.7).

Proof Hannan (1973), Fox and Taqqu (1986) (also see Beran 1986), Giraitis and Surgailis (1990), Horváth and Shao (1999).

Remark 10.4 A further approximation follows from a Riemann sum approximation of W_n using the periodogram $I(\lambda)$ at Fourier frequencies $\lambda_j = 2\pi j/n$:

$$\mathcal{L}_{n,\text{Whittle}}(\theta) \approx \tilde{\mathcal{L}}_{n,\text{Whittle}}(\theta) = \log\sigma_\varepsilon^2 + \frac{2}{n}\sum_{j=1}^{N_n}\frac{I(\lambda_j)}{f_X(\lambda_j;\theta)}.$$

10.3 Autoregressive Approximation

Definition 10.3 Let

$$\theta = \left(\sigma_{\varepsilon}^{2}, \eta\right) \in \Theta \subseteq \mathbb{R}^{p+1}$$

$$f_X(\lambda; \theta)\ (\lambda \in [-\pi, \pi], \theta \in \Theta) \text{ family of spectral densities}$$

as in Definition 10.1, and

$$X_t = \sum_{j=0}^{\infty} a_j\left(\eta^0\right)\varepsilon_{t-j} = A(B)\,\varepsilon_t\ (t \in \mathbb{Z}, \theta^0 = \left(\sigma_{\varepsilon}^{2}, \eta\right) \in \Theta^0 \subseteq \mathbb{R}^{p+1})$$

as in Definition 10.1. Moreover, assume

$$\varepsilon_t = \varepsilon_t\left(\eta^0\right) = \sum_{j=0}^{\infty} b_j\left(\eta^0\right) X_{t-j},$$

and define

$$\varepsilon_t(\eta) = \sum_{j=0}^{\infty} b_j(\eta)\, X_{t-j}$$

and

$$S_n(\eta) = \sum_{t=1}^{n} \varepsilon_t^2(\eta).$$

If

$$F_{\varepsilon,1}(x) = \Phi(x),$$

then $\hat{\theta} = \hat{\theta}_{AR}$ defined by

$$\hat{\eta} = \underset{\eta}{\operatorname{argmin}}\, S_n(\eta) \tag{10.16}$$

and

$$\hat{\sigma}_{\varepsilon}^{2} = \frac{1}{n} S_n(\hat{\eta})$$

is called the maximum likelihood estimator, or MLE, of θ^0, based on the autoregressive representation. If

$$F_{\varepsilon,1}(x) = \Phi(x),$$

then $\hat{\theta}_{AR}$ is called quasi maximum likelihood estimator, or QMLE, of θ^0, based on the autoregressive representation.

Remark 10.5 Under suitable regularity conditions, (10.16) is equivalent to

$$\Psi_\varepsilon(\hat{\eta}) = 0$$

where

$$\Psi_\varepsilon(\eta) = \frac{\partial}{\partial \eta} S_n(\eta) = \sum_{t=1}^{n} \varepsilon_t(\eta) \frac{\partial}{\partial \eta} \varepsilon_t(\eta).$$

Problem 10.4

$$\varepsilon_t(\eta) = \sum_{j=0}^{\infty} b_j(\eta) X_{t-j}$$

contains unobserved values $X_0, X_{-1}, \ldots$

Solution 10.4 For t large enough,

$$\varepsilon_t(\eta) \approx e_t(\eta) = \sum_{j=0}^{t-1} b_j(\eta) X_{t-j}. \tag{10.17}$$

Hence, define the approximate estimator $\hat{\theta}_{AR}$ by

$$\Psi_e(\hat{\eta}_{AR}) = 0, \hat{\sigma}_\varepsilon^2 = \frac{1}{n} \sum_{t=2}^{n} e_t^2(\hat{\eta}_{AR}) \tag{10.18}$$

where

$$\Psi_e(\eta) = \sum_{t=1}^{n-1} e_t(\eta) \frac{\partial}{\partial \eta} e_t(\eta).$$

Theorem 10.3 *Under the assumptions of Theorem 10.1 and additional regularity assumptions,*

$$\sqrt{n}\left(\hat{\theta}_{AR} - \theta^0\right) \underset{d}{\rightarrow} N(0, \Sigma_{\text{MLE}})$$

where Σ_{MLE} is given in (10.7).

Proof See e.g. Beran (1995), Velasco and Robinson (2000).

10.4 Model Choice

Problem 10.5

$$X_t \ (t \in \mathbb{Z}) \text{ time series model} \tag{10.19}$$

$$\Omega = \left(\mathbb{R}^k\right)^T, \mathscr{F} = \sigma\text{-algebra on } \Omega \tag{10.20}$$

$$X_t \sim P_{\theta^0}$$

$$P_\theta \ (\theta \in \Theta) \text{ family of probability distributions on } (\Omega, \mathscr{F}) \tag{10.21}$$

$$\theta^0 \in \Theta = \bigcup_{m \geq 0} \Theta_m, \ \dim \Theta_m = m + 1 \tag{10.22}$$

$$\theta^0 \text{ and } m \text{ unknown} \tag{10.23}$$

$$\Rightarrow \text{estimation of } m?$$

Solution 10.5 Model choice criteria based on information theory.

Remark 10.6 In addition to formal model choice criteria one may apply heuristic diagnostic tools, such as checking independence of $\varepsilon_t(\hat{\eta})$, empirical acf and partial acf etc. These are not discussed here.

Definition 10.4 Let

$$X_t \ (t \in \mathbb{Z}) \text{ time series model } P_{\theta^0} \text{ as in (10.20)–(10.23)},$$

$$Y_t \ (t \in \mathbb{Z}) \text{ independent of } X_t \ (t \in \mathbb{Z}), \ (Y_t)_{t \in \mathbb{Z}} \sim P_{\theta^0},$$

$$\hat{\theta}\,(X, m) = \underset{\theta \in \Theta_m}{\operatorname{argmax}} \mathscr{L}_n\,(\theta; X_1, \ldots, X_n)\,.$$

Define the loss function

$$L\left(m; \hat{\theta}\,(X, m)\right) = -2E_{y,\theta^0}\left[\mathscr{L}_n\left(\hat{\theta}\,(X, m)\,; Y_1, \ldots, Y_n\right)\right]$$

where $\mathscr{L}_n(\hat{\theta}(X, m); Y_1, \ldots, Y_n)$ is the loglikelihood function evaluated at $Y_1, \ldots, Y_n$ and the (now fixed) parameter $\hat{\vartheta}(X, m)$, and the expected value $E_{y,\vartheta^0}[\cdot]$ is taken with respect to the distribution of $Y_1, \ldots, Y_n$ which is specified by the correct parameter θ^0. Moreover, define the risk function

$$R(m) = E_{x,\theta_0}\left[L\left(m; \hat{\theta}(X, m)\right)\right]$$

where the expectation $E_{x,\vartheta^0}[\cdot]$ is taken with respect to the distribution of $X_1, \ldots, X_n$.

Theorem 10.4 *Assumptions:*

$$X_t \ (t \in \mathbb{Z}) \text{ time series model } P_{\theta^0} \text{ as in (10.20)–(10.23)},$$

$$\hat{\theta}(X, m) = \left[\hat{\sigma}_\varepsilon^2(X, m), \hat{\eta}(X, m)\right] = \underset{\theta \in \Theta_m}{\arg\max} \mathcal{L}_n(\theta; X_1, \ldots, X_n).$$

Then, under suitable additional regularity conditions,

$$R(m) = n \log \hat{\sigma}_\varepsilon^2(X, m) + 2m + C\left(n, \theta^0\right) + o(1)$$

where C does not depend on m and $\hat{\theta}(X, m)$.

Remark 10.7 C is irrelevant for choosing m.

Definition 10.5

$$X_t \ (t \in \mathbb{Z}) \text{ time series model } P_{\theta^0} \text{ as in (10.20)–(10.23)},$$

$$\theta^0 \in \Theta_{m_0}$$

$$\hat{\theta}(X, m) = \left[\hat{\sigma}_\varepsilon^2(X, m), \hat{\eta}(X, m)\right] = \underset{\theta \in \Theta_m}{\operatorname{argmax}} \mathcal{L}_n(\theta; X_1, \ldots, X_n).$$

Then,

$$AIC(m) = n \log \hat{\sigma}_\varepsilon^2(X, m) + 2m$$

(Akaike 1973, 1974), and, for a given $m_{\max} \geq 0$,

$$\hat{m}_{AIC} = \underset{0 \leq m \leq m_{\max}}{\operatorname{argmin}} AIC(m)$$

is an estimator of m_0 according to the Akaike Information Criterion (AIC).

Under general assumptions one can show

$$P(\hat{m}_{AIC} < m_0) \to 0, P(\hat{m}_{AIC} > m_0) \to c \in (0, 1). \tag{10.24}$$

To avoid overestimation of m_0, Schwarz (1978) introduced the so-called BIC (Bayesian Information Criterion)

$$BIC\,(m) = n \log \hat{\sigma}_{\varepsilon}^2 + 2m \log n,$$

and Hannan and Quinn (1979) proposed the HIC

$$HIC\,(m) = n \log \hat{\sigma}_{\varepsilon}^2 + 2m \cdot c \log \log n$$

where $c > 1$. For both criteria one can show

$$\lim_{n \to \infty} P\,(\hat{m} = m_0) = 1$$

(see also e.g. Shibata 1980, 1981, Beran et al. 1998, McQuarrie and Tsai 1998, Claeskens and Hjort 2008 and references therein).

References

Adenstedt, R. K. (1974). On large-sample estimation for the mean of a stationary random sequence. *The Annals of Statistics, 2*(6), 1095–1107.

Akaike, H. (1973). Information theory and an extension of the maximum likelihood principle. In B. N. Petrov (Hrsg.) & F. Csáki (Eds.), *Proceedings of the Second International Symposium on Information Theory* (pp. 267–281). Budapest: Akademiai Kiado.

Akaike, H. (1974). A new look at the statistical model identification. *IEEE Transactions on Automatic Control, 19*(6), 716–723.

Albiac, F., & Kalton, N. J. (2016). *Topics in Banach space theory*. New York: Springer.

Anderson, T. W. (1971). *The statistical analysis of time series*. New York: Wiley.

Andrews, G. E., Askey, R., & Roy, R. (1999). *Special functions*. Cambridge: Cambridge University Press.

Ash, R. B. (1972). *Real analysis and probability*. New York: Academic Press.

Basrak, B., Davis, R. A., & Mikosch, T. (1999). The sample ACF of a simple bilinear process. *Stochastic Processes and Their Applications, 83*, 1–14.

Beltrao, K. I., & Bloomfield, P. (1987). Determining the bandwidth of a kernel spectrum estimate. *Journal of Time Series Analysis, 8*, 21–38.

Beran, J. (1986). *Estimation, testing and prediction for self-similar and related processes*. Ph.D. Thesis, ETH Zürich.

Beran, J. (1994). *Statistics for long-memory processes*. New York: Chapman & Hall/CRC Press.

Beran, J. (1995). Maximum likelihood estimation of the differencing parameter for invertible short and long-memory ARIMA models. *Journal of the Royal Statistical Society, Series B, 57*, 659–672.

Beran, J., Bhansali, R. J., & Ocker, D. (1998). On unified model selection for stationary and nonstationary short- and long-memory autoregressive processes. *Biometrika, 85*(4), 921–934.

Beran, J., Feng, Y., Ghosh, S., & Kulik, R. (2013). *Long-memory processes*. New York: Springer.

Beran, J., & Künsch, H. (1985). *Location estimators for processes with long-range dependence*. Research Report No. 40, Seminar für Statistik, ETH, Zurich.

Berkes, I., Horváth, L., & Kokoszka, P. (2003). GARCH processes: Structure and estimation. *Bernoulli, 9*, 201–227.

Berkes, I., Horváth, L., & Kokoszka, P. (2004). Probabilistic and statistical properties of GARCH processes. *Fields Institute Communications, 44*, 409–429.

Bhattacharya, R., & Waymire, E. C. (2007). *A basic course in probability theory*. New York: Springer.

Birkhoff, G. D. (1931). Proof of the ergodic theorem. *Proceedings of the National Academy of Sciences USA, 17*, 656–660.

J. Beran, *Mathematical Foundations of Time Series Analysis*,
https://doi.org/10.1007/978-3-319-74380-6

Bollerslev, T. (1986). Generalized autoregressive conditional heteroskedasticity. *Journal of Econometrics, 31*, 307–327.

Bougerol, P., & Picard, N. (1992a). Strict stationarity of generalized autoregressive processes. *The Annals of Probability, 20*(4), 1714–1730.

Bougerol, P., & Picard, N. (1992b). Stationarity of GARCH processes and of some nonnegative time series. *Journal of Econometrics, 52*, 115–127.

Box, G. E. P., & Jenkins, G. M. (1970). *Time series analysis: Forecasting and control*. San Francisco: Holden-Day.

Bradley, R. C. (2007). *Introduction to strong mixing conditions* (Vols. 1, 2, and 3). Heber City, UT: Kendrick Press.

Brandt, A. (1986). The stochastic equation $Y_{n+1} = A_n Y_n + B_n$ with stationary coefficients. *Advances in Applied Probability, 18*, 211–220.

Breiman, L. (1968). *Probability*. Redding, MA: Addison-Wesley.

Breiman, L. (1992). *Probability*. Philadelphia: SIAM.

Brillinger, D. R. (1969). Asymptotic properties of spectral estimates of second order. *Biometrika, 56*, 375–390.

Brillinger, D. R. (2001). *Time series: Data analysis and theory*. Philadelphia: Society for Industrial and Applied Mathematics.

Brockwell, P. J., & Cline, D. B. H. (1985). Linear prediction of ARMA processes with infinite variance. *Stochastic Processes and Applications, 19*, 281–296.

Brockwell, P. J., & Davis, R. A. (1991). *Time series: Theory and methods*. New York: Springer.

Bruckner, A. M., Bruckner, J. B., & Thomson, B. S. (1997). *Real analysis*. Upper Saddle River, NJ: Prentice-Hall.

Caines, P. E. (1988). *Linear stochastic systems*. New York: Wiley.

Cantelli, F. P. (1933). Sulla determinazione empirica delle leggi di probabilità. *Giornale dell'Istituto Italiano degli Attuari, 4*, 421–424.

Chatfield, C. (2003). *The analysis of time series: An introduction* (6th ed.). Boca Raton: CRC Press.

Claeskens, G., & Hjort, N. L. (2008). *Model selection and model averaging*. Cambridge: Cambridge University Press.

Dahlhaus, R. (1989). Efficient parameter estimation for self-similar processes. *The Annals of Statistics, 17*(4), 1749–1766.

Dahlhaus, R. (1995). Efficient location and regression estimation for long range dependent regression models. *The Annals of Statistics, 23*, 1029–1047.

Dhrymes, P. (1997). *Time series, unit roots and cointegration*. Bingley, UK: Emerald Group Publishing Limited.

Dickey, D. A., & Fuller, W. A. (1979). Distribution of the estimators for autoregressive time series with a unit root. *Journal of American Statistical Association, 74*, 427–431.

Diggle, P. (1996). *Time series: A biostatistical introduction*. Oxford: Oxford University Press.

Doob, J. L. (1953). *Stochastic processes*. New York: Wiley.

Doukhan, P. (1995). *Mixing: Properties and examples*. New York: Springer-Verlag.

Durbin, J., & Koopmann, S. J. (2012). *Time series analysis by state space methods*. Oxford: Oxford University Press.

Duren, P. (1970). *Theory of H_p-spaces*. New York: Academic Press.

Edwards, R. E. (1967). *Fourier series: A modern introduction* (Vols. I, II). New York: Holt, Rinehart and Winston.

Engle, R. F. (1982). Autoregressive conditional heteroscedasticity with estimates of variance of United Kingdom inflation. *Econometrica, 50*(4), 987–1008.

Engle, R. F., & Granger, C. W. J. (1987). Co-integration and error correction: Representation, estimation, and testing. *Econometrica, 55*, 251–276.

Fox, R., & Taqqu, M. S. (1986). Large-sample properties of parameter estimates for strongly dependent stationary Gaussian time series. *The Annals of Statistics, 14*, 517–532.

Franke, J., & Härdle, W. (1992). On bootstrapping kernel spectral estimates. *The Annals of Statistics, 20*(1), 121–145.

Fuller, W. A. (1976). *Introduction to statistical time series*. New York: Wiley.

Furstenberg, H., & Kesten, H. (1960). Products of random matrices. *Annals of Mathematical Statistics, 31*, 457–469.
Ghosh, S. (2017). *Kernel smoothing. Principles, methods and applications*. New York: Wiley.
Gihman, I. L., & Skorohod, A. V. (1974). *The theory of stochastic processes I* (S. Kotz Trans.). Berlin: Springer-Verlag.
Giraitis, L., Kokoszka, P., & Leipus, R. (2000). Stationary ARCH models: Dependence structure and central limit theorem. *Econometric Theory, 16*, 3–22.
Giraitis, L., Koul, H. L., & Surgailis, D. (2012). *Large sample inference for long memory processes*. London: Imperial College Press.
Giraitis, L., & Surgailis, D. (1990). A central limit theorem for quadratic forms in strongly dependent linear variables and its application to asymptotical normality of Whittle's estimate. *Probability Theory and Related Fields, 86*(1), 87–104.
Glivenko, V. (1933). Sulla determinazione empirica delle leggi di probabilità. *Giorn. Ist. Ital. Attuari, 4*, 92–99.
Gradshteyn, I. S., & Ryzhik, I. M. (1965). *Table of integrals, series, and products*. New York: Academic Press.
Granger, C. W. J., & Newbold, P. (1974). Spurious regressions in econometrics. *Journal of Econometrics, 2*(2), 111–120.
Grenander, U., & Rosenblatt, M. (1957). *Statistical analysis of stationary time series*. New York: Wiley.
Grenander, U., & Szegö, G. (1958). *Toeplitz forms and their application*. Berkeley: University of California Press.
Hamilton, J. D. (1994). *Time series analysis*. Princeton: Princeton University Press.
Hannan, E. J. (1970). *Multiple time series*. New York: Wiley.
Hannan, E. J. (1973). Central limit theorems for time series regression. *Zeitschrift für Wahrscheinlichkeitstheorie und Verwandte Gebiete, 26*, 157–170.
Hannan, E. J., & Kanter, M. (1977). Autoregressive processes with infinite variance. *Journal of Applied Probability, 14*, 411–415.
Hannan, E. J., & Quinn, B. G. (1979). The determination of the order of an autoregression. *Journal of the Royal Statistical Society, Series B, 41*(2), 190–195.
Horváth, L., & Shao, Q.-M. (1999). Limit theorems for quadratic forms with applications to Whittle's estimate. *The Annals of Applied Probability, 9*(1), 146–187.
Hosoya, Y. (1997). A limit theory for long-range dependence and statistical inference on related models. *The Annals of Statistics, 25*(1), 105–137.
Johansen, S. (1995). *Likelihood-based inference in cointegrated vector autoregressive models*. Oxford: Oxford University Press.
Johnsonbaugh, R. F., & Pfaffenberger, W. E. (2010). *Foundations of mathematical analysis*. Mineola, NY: Dover Publications.
Kallenberg, O. (2002). *Foundations of modern probability*. New York: Springer.
Katznelson, Y. (1976). *An introduction to harmonic analysis*. Cambridge: Cambridge University Press.
Kazakevičius, V., & Leipus, R. (2002). On stationarity of the ARCH(∞) model. *Econometric Theory, 18*(1), 1–16.
Kingman, J. F. C. (1973). Subadditive ergodic theory. *The Annals of Probability, 1*, 883–909.
Kolmogorov, A. N. (1941a). Stationary sequences in Hilbert space. *Bull. Moscow State U. Math., 2*, 40 pp.
Kolmogorov, A. N. (1941b). Interpolation and extrapolation of stationary random sequences. *Izv. Akad. Nauk SSSR, Ser. Mat., 5*, 3–14.
Kolmogorov, A. N. (1956). *Foundations of the theory of probability* (2nd ed.). New York: Chelsea Publishing Company.
Koopmans, L. H. (1974). *The spectral analysis of time series*. New York: Academic Press.
Koosis, P. (1980). *Introduction to H_p-spaces*. Cambridge: Cambridge University Press.
Lang, S. (2003). *Complex analysis* (4th ed.). New York: Springer.

Lighthill, M. J. (1962). *Introduction to Fourier analysis and generalised functions*. Cambridge monographs on mechanics and applied mathematics. Cambridge University Press: Cambridge.

Liu, W., & Wu, W. B. (2010). Asymptotics of spectral density estimates. *Econometric Theory, 26*, 1218–1245.

Lütkepohl, H. (2006). New introduction to multiple time series analysis. Berlin: Springer.

McKean, H. (2014). *Probability: The classical limit theorems*. Cambridge: Cambridge University Press.

McQuarrie, A. D., & Tsai, C. L. (1998). *Regression and time series model selection*. Singapore: World Scientific.

Mikosch, T., & Starica, C. (2000). *Limit theory for the sample autocorrelations and extremes of a GARCH (1,1) process*. *Annals of Statistics, 28*(5), 1427–1451.

Nelson, D. B. (1990). Stationarity and persistence in the GARCH(1,1) model. *Econometric Theory, 6*, 318–334.

Nyquist, H. (1928). Certain topics in telegraph transmission theory. *Transactions of the American Institute of Electrical Engineers, 47*, 617–644.

Parzen, E., Tanabe, K., & Kitagawa, G. (Eds.) (1998). *Selected papers of Hirotugu Akaike*. New York: Springer.

Phillips, P. C. B. (1986). Understanding spurious regressions in econometrics. *Journal of Econometrics, 33*(3), 311–340.

Priestley, M. B. (1981). *Spectral analysis and time series*. San Diego: Academic Press.

Rosenblatt, M. (1956). A central limit theorem and a strong mixing condition. *Proceedings of the National Academy of Sciences USA, 42*, 43–47.

Rosenblatt, M. (1984). Asymptotic normality, strong mixing, and spectral density estimates. *Annals of Probability, 12*, 1167–1180.

Rynne, B. P., & Youngson, M. A. (2007). *Linear functional analysis*. London: Springer.

Said, E., & Dickey, D. A. (1984). Testing for unit roots in autoregressive moving average models of unknown order. *Biometrika, 71*, 599–607.

Samarov, A., & Taqqu, M. S. (1988). On the efficiency of the sample mean in long-memory noise. *Journal of Time Series Analysis, 9*(2), 191–200.

Schwarz, G. (1978). Estimating the dimension of a model. *Annals of Statistics, 2*(6), 461–464.

Shannon, C. E. (1948). A mathematical theory of communication. *Bell Systems Technical Journal, 27*, 379–423.

Shao, X., & Wu, W. B. (2007). Asymptotic spectral theory for nonlinear time series. *The Annals of Statistics, 35*(4), 1773–1801.

Shibata, R. (1980). Asymptotically efficient selection of the order of the model for estimating parameters of a linear process. *The Annals of Statistics, 8*, 147–164.

Shibata, R. (1981). An optimal autoregressive spectral estimate. *The Annals of Statistics, 9*, 300–306.

Shiryaev, A. N. (1995). *Probability* (2nd ed.). New York: Springer.

Shumway, R. H., & Stoffer, D. S. (2017). *Time series analysis and its applications: With R examples* (4th ed.). New York: Springer.

Stieltjes, T. J. (1894). Recherches sur les fractions continues. *Annales de la Faculté des Sciences Toulouse, VIII*, 1–122.

Stout, H. F. (1974). *Almost sure convergence*. New York: Academic Press.

Tong, H. (1990). *Non-linear time series: A dynamical system approach*. Oxford: Oxford University Press.

van der Vaart, A. W. (1998). *Asymptotic statistics*. Cambridge: Cambridge University Press.

Velasco, C., & Robinson, P. M. (2000). Whittle pseudo-maximum likelihood estimation for nonstationary time series. *Journal of the American Statistical Association, 95*(452), 1229–1243.

Walters, P. (1982). *An introduction to ergodic theory*. New York: Springer.

Whittle, P. (1962). Gaussian estimation in stationary time series. *Bulletin de L'Institut International de Statistique, 39*, 105–129.

Wiener, N., & Masani, P. (1958). Prediction theory of multivariate stochastic processes, II. *Acta Mathematica, 99*, 93–137.

Woodward, W.A., Gray, H. L., & Elliott, A. C. (2016). *Applied time series analysis with R*. Boca Raton: CRC Press.

Yajima, Y. (1985). On estimation of long-memory time series models. *Australian Journal of Statistics, 27*(3), 303–320.

Yajima, Y. (1988). On estimation of a regression model with long term errors. *The Annals of Statistics, 16*(2), 791–807.

Yajima, Y. (1991). Asymptotic properties of the LSE in a regression model with long-memory stationary errors. *The Annals of Statistics, 19*, 158–177.

Young, N. (1988). *An introduction to Hilbert space*. Cambridge: Cambridge University Press.

Zurbenko, I. G. (1986). *The spectral analysis of time series*. Amsterdam: North-Holland.

Zygmund, A. (1968). *Trigonometric series* (Vol. 1). Cambridge: Cambridge University Press.

Author Index

J. Beran, *Mathematical Foundations of Time Series Analysis*,
https://doi.org/10.1007/978-3-319-74380-6

Subject Index

J. Beran, *Mathematical Foundations of Time Series Analysis*,
https://doi.org/10.1007/978-3-319-74380-6

MIX
Papier aus verantwortungsvollen Quellen
Paper from responsible sources
FSC® C105338

If you have any concerns about our products,
you can contact us on
ProductSafety@springernature.com

In case Publisher is established outside the EU,
the EU authorized representative is:
Springer Nature Customer Service Center GmbH
Europaplatz 3, 69115 Heidelberg, Germany

Printed by Libri Plureos GmbH
in Hamburg, Germany